Rebecca and Drew,

I'm delighted to share my first cookbook with you. My easy-to-understand science tips will enhance your everyday cooking and advance your culinary adventures. You will be able to make great food again and again and again. It's not magic. It's science. It's like Magic!

Have fun! Bon appétit,

Colleen

Like Magic!

Use the magic of science to release the chef in you

COLLEEN HISCOCK, B. Sc.
PHOTOGRAPHY BY SALLY LEDREW

Colleen Hiscock
PO Box 149
Rocky Harbour NL A0K 4N0
Email: colleen.hiscock@gmail.com
Website: www.chefcolleen.ca
Instagram: @chefcolleenhiscock
TikTok: @chefcolleenh
YouTube Channel: @chefcolleenhiscock

Like Magic!
Copyright © 2022 by Colleen Hiscock

All rights reserved. No part of this publication may be reproduced, distributed, or transmitted in any form or by any means, including photocopying, recording, or other electronic or mechanical methods, without the prior written permission of the author, except in the case of brief quotations embodied in critical reviews and certain other non-commercial uses permitted by copyright law.

tellwell

Tellwell Talent
www.tellwell.ca

ISBN
978-0-2288-7395-2 (Hardcover)
978-0-2288-7394-5 (Paperback)

Dedicated to women who excel in the field of science, technology, engineering and mathematics (STEM) and love their own ideas and have the drive to make them come alive.

To my daughters, Erica (BA, BN, RN, CEN), and Heather (BSc., MSc.). I am so proud of the who you are; strong, smart, beautiful women making great and significant contributions in your own fields.

To my husband, Leslie who listens to my many crazy ideas and with whom I am free to run, be myself and shoot for the stars.

To my mother, Judy, who is still a strong, optimistic role model.

To my father, Harold, who is such a positive human being even in the face of adversity.

To my late grandmother, whose presence and pride I still feel.

Here's to all of you!

By denying scientific principles, one may maintain any paradox.
-Galileo Galilei

PRAISE FOR
Like Magic!

Colleen's exuberance and passion shine through in her cookbook. I love the way in which she takes the time to simplify and explain the science behind cooking. Her warmth comes through in the lovely stories she tells and the experiences that she shares. The YouTube video tutorials, and the "Rule of Thumb" boxes are a great addition to the book. I look forward to trying some of her favourite recipes and adding them to my repertoire.

Jonathan Preskow
Executive Chef and Owner, JP Fine Foods
COO, Great North Smokehouse

I met Colleen during one of my many encounters as a culinary consultant with various food and beverage companies. She had a great reputation in the industry for your knowledge of food science and product development. Cooking is a marriage of craft and science and Colleen has managed to bridge the two beautifully in this cookbook, due to her wealth of knowledge and experience in both fields. I have a deep passion for the science behind cooking for decades now and having just finished reviewing this cookbook, I was struck by how different it is to so many cookbooks I've read. Not only are the photos beautiful and the recipes easy to follow but I love her "rule of thumb" throughout the book, the tips with the recipes, and the "How to" sections with so much valuable knowledge which tells you why you do the things you do in cooking. And the QR codes for selected recipes are a nice touch so you can view the "How To" on YouTube. I'm sure you'll love this book as much as I do.

John Placko
Chef and Owner, Modern Culinary Academy and KOCO Chocolate Company

Colleen has shared her magic wand with us! 'Like Magic' is an approachable, easy to follow book that takes her food science wisdom and serves it up for us to learn! This is a must have book for your kitchen that you will learn techniques you can use over and over again. It makes information that can often come across stale and uninteresting sparkle through her powerful easy to understand visuals and antidotes. The meat section is next level. Having known Colleen for some time I can tell you that her voice in this book is exactly what she sounds like when she teaches you something. You will want her as your new friend by the end of this book!

David Gosine
Food and Wine Writer and Consultant

I'm very excited about Colleen's decision to create this wonderful book. Her understanding of the science of cooking combined with her bursting motivation and contagious positive energy would drive anyone to want to start cooking. Being an avid self- taught cook, I'm delighted to have so many of her delicious recipes and ideas in one place.

Michelle Doyle
Singer, Actor, Chef and Owner, Doyle's Daily Dish

Colleen's passion for food is clearly reflected in the beautiful pages of this book. In her heart she is a teacher and uses these pages to explain the "why" of cooking, the science behind the magic, which will help us to be more successful in every recipe, not just the ones here. The meat science she taught me has resulted in my reputation for amazing burgers. She deserves the credit, but I keep it all!

Kasia Luckevich
Chef and Director, Research and Development, Sofina Foods Inc.

TABLE OF CONTENTS

1. Introduction 1
    My Story 1
2. Get Ready 4
    My Story: Key to Success 4
    The Magic of Science: Preparation 5
3. The Bakery 19
    My Story: Company's Coming 19
    The Magic of Science: Leavening and Browning 19
4. Building Blocks for Salads 35
    My Story: Eating Salad 35
    The Magic of Science: Emulsions 35
5. Stocks, Soups and Sauces 63
    My Story: Making of a Seafood Chowder 63
    The Magic of Science: Thickening 64
6. Wow Potatoes and Veggie Side Dishes 97
    My Story: Size for speed 97
    The Magic of Science: Size Matters 97
    WOW! Potatoes 99
7. Get Crackin' with Eggs 123
    My Story: Flawless Crème Brûlée 123
    The Magic of Science: Egg Gel Networks 124
    The Functionality of Eggs 125
8. TA DA! Fish and Seafood 151
    My Story: Seeing is Believing 151
    The Magic of Science: Fish is Not Smelly 152
    Cooking Fish and Seafood 153

9. Magic in Meat 172
    Meat and its Composition 172

10. Turning Up the Heat on Meat 177
    My Story: Touting Two Ts 177
    The Magic of Science: Heat Transfer 178
    Cooking Meat 179

11. Bye Bye, Dry Birdie 197
    My Story: Burger Magic 197
    The Magic of Science: Water-Holding Capabilities 198

12. Nutritious and Sweet 219
    My Story: Sharing 219
    The Magic of Science: Fruit and Oatmeal 220

Appendix A 239
Glossary 241
References 245
Index 247
About the Author 253
Acknowledgements 255

1. Introduction

MY STORY

Everyone has a relationship with food in some way, shape, or form. It's an essential part of our daily lives, and for many, if we're not eating, we're thinking about eating and what we'll be having for supper. Everyone has a favourite food as well as food that turns their stomach, and yet no two palates are the same. You ask ten people what they'd like to eat if they could have anything, and you'll get ten different menus. It is a distinguishing characteristic of global and even local cultural differences.

Many enjoy creating and presenting a signature dish to their family and guests. Often what we put together, how we did it, and the special little trick that created this perfect food is what all enjoy and marvel at. The careful selection of the ingredients and tools and the precise technique used to create the exact smell and how good it looks and tastes are what makes a dish the centre of the dinner conversation before, during, and after the meal. It becomes that memorable experience that is shared and lingers on long after the event has passed. For most, breaking away from their usual repertoire to try a new dish or two can be daunting. So this cookbook uses recipes that are familiar and reinvents them or even upgrades them to keep the traditional repertoire fresh and vibrant.

More people are impressed with the designation of chef than food scientist, but my science-based foundation cements my success at making perfect food one plate at a time, for large groups or en masse for the population. I have spent thirty years in the food industry focused on developing a variety of food products: food for the masses, food for value sought, and healthful, wholesome food made cheaply, quickly, effectively—all backed by scientific principles of biology, chemistry, physics, and math. In recent years I earned Red Seal certification in the culinary field and am called "Chef" these days.

Here, for this cookbook, *Like Magic!*, I explain in simple terms the fundamentals of food science for the home cook and foodie—those who are interested in cooking and those who want to improve their cooking by solving very common problems. For example, issues with hollandaise sauce breaking, inexpensive cuts of meat being very tough and chewy, or turkey that is too dry often lead to grievances from friends and followers. Many recipes and dishes ask for extra, seemingly functional ingredients to bandage a common issue such as holding a burger together on the BBQ by using eggs and breadcrumbs. However, scientific principles can easily be used to solve frustrations while creating better flavour, reducing cost, and improving those signature recipes.

I'm a Newfoundlander, and we are known for storytelling, so each recipe and science principle begins with a little story of why it was chosen; it could be to dispel the myth of a small-town community, or to enhance and simplify a current practice, or just a discussion to share some simple product knowledge used in the food industry. Additionally, my heritage and upbringing were warm and welcoming, and I want the book to reflect that and be approachable and helpful to everyone, both science-minded folks and those who are interested in cooking. It is not meant to be a nerdy, sturdy book, best used by the few and a doorstop for most, though it includes some science and math principles—the ones we all wondered about how they could possibly be useful when studying them in grade school. *Like Magic!* contains beautiful photographs of food, bridging deliciousness with the science that makes it so. It contains interesting anecdotes of the chosen recipe along with the science connector of "Why it works" as well as two to three extension recipe options to further demonstrate the use of the scientific guidelines. If you follow the guidelines, you can create your own signature recipe confidently and take it further if you wish. My overall objective is to give you practical guidance for cooking that will grow your confidence that your dish will taste great, look great, and smell great every time you make it. Unfailing, science based, delicious cooking.

Science does not know its debt to imagination.
-Ralph Waldo Emerson

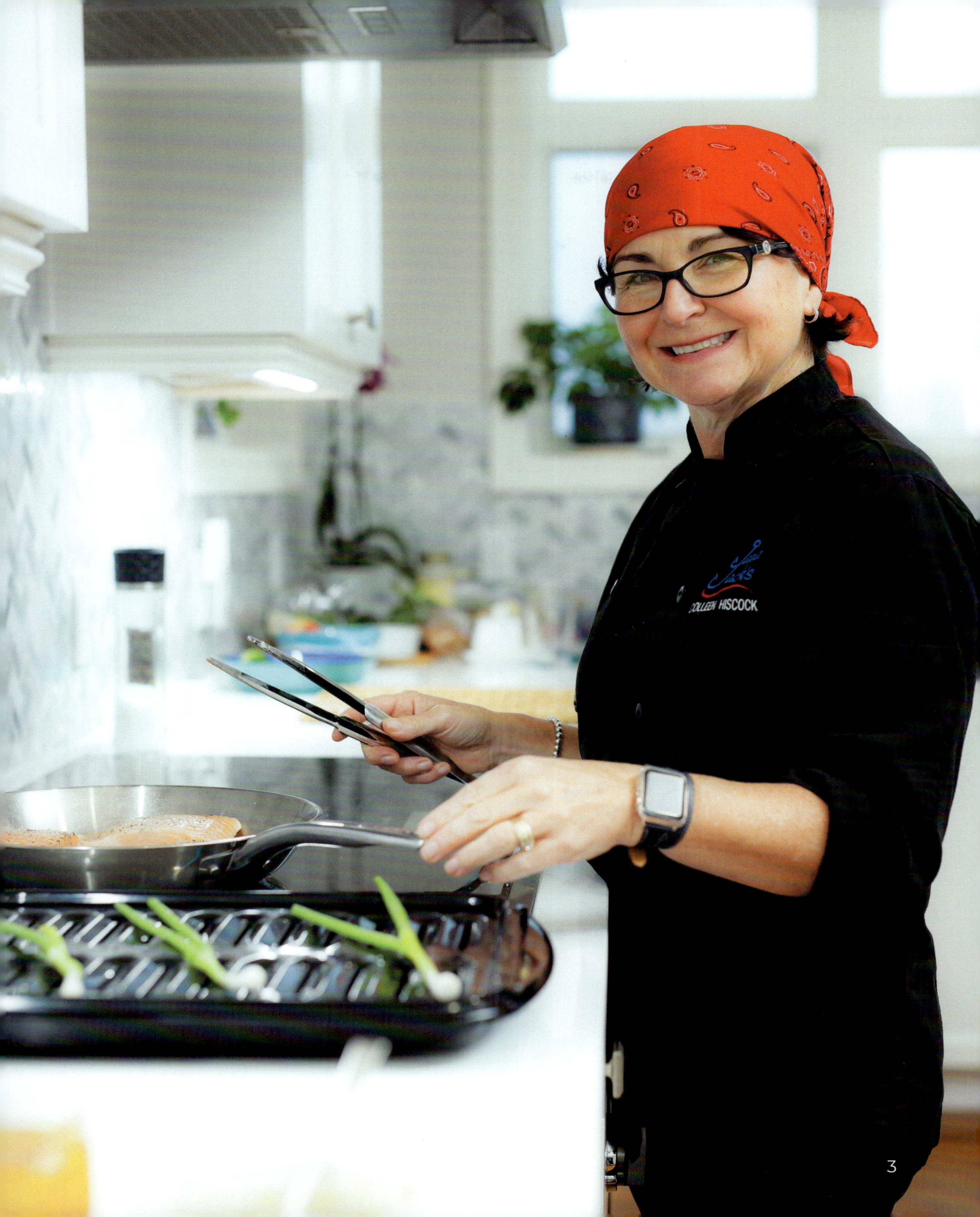
COLLEEN HISCOCK

2. Get Ready

MY STORY: KEY TO SUCCESS

I'm a math advocate. I think math is everywhere and in everything. From simple addition and fractions to the area under a curve, math is in everyday life. I am often surprised when people can't see where the application of math is because it is all around us. I also believe anyone can learn math, like a language or a tool. Math is a powerful tool. I count everything, portion everything, and measure time, yield, and waste before I start to cook, during the process, and afterward. I *always* think success, particularly in cooking, comes down to preparation and knowing my preparation is precise and complete. This is contrary to most culinary professionals and true if you visualize cooking as the magic created by the Swedish chef from *The Muppet Show*. My staff (and my children) have learned that relentless up-front thinking, research, understanding, and premeasurement to control what is controllable ensure success whether you're in a customer presentation or product launch, hosting Christmas dinner, or just making dinner quickly. Perhaps it's that my last name was Murphy that set me off this way. Perhaps "getting ready" is an inherent trait seen by some as a flaw, and yet for successful manufacturers using Six Sigma principles, it is one of a few pinnacle tenets that is key to success. In every instance, getting ready and being ready is what drives confidence and success. So let's get ready!

THE MAGIC OF SCIENCE: PREPARATION

Success is multifaceted. It is not perfect tools, perfect measurements, the best ingredients, the best technique, everything at the ready. It's a combination of them all. A good frying pan, spatula, and wooden spoon along with any bowl, a whisk, and two eggs prepared by whisking vigorously, then stirring gently, result in the most tender curds of simply exquisite, scrambled eggs. So I share some details about why measurement, tools, and organization can be the key to enjoyment in preparing and eating the dish you want to make.

- Tools I care about
- Measurement, portions, and cooking math
- Mise en place (getting organized)

TOOLS I CARE ABOUT

Here are a few of my most used tools. While there are whole industries out there that sell you on the "best" or "top" quality version of every kitchen gadget and tool required to make the best food, my experience has shown me I can make do with whatever gadget is at my fingertips! Indeed, I would argue the list of important, effective, specific tools includes only a few, while other tools are either sufficient or are simply nice to have.

If you're working with a budget for your culinary tools and want to know what to invest in, here is my short list.

KNIVES

While I do have other knives I picked up here and there, such as a serrated knife, turning knife, trimming knife, and tomato knife, I can get along with an inexpensive, nonspecific version of these. Instead I recommend you invest in just these four knives and a steel.

- Chef's knife: Your multipurpose knife. It has enough weight to push through joints, enough of a blade to slice large cuts of meat or carve, and a pointy tip to cut onions and dice them. This knife's shape facilitates the rocking motion of mincing and fine chopping specifically.
- Santoku knife: Another multipurpose knife used for cutting, chopping, dicing, and slicing vegetables. This knife has a bevelled edge that releases the sticky moisture vegetables release after each downward slice motion.

- Paring knife: A small knife used for small jobs and more intricate work. I have a few of these on hand and don't spend too much money on them because I tend to lose them, drop them, or just wear them out. I can purchase many inexpensive ones for the same price of a "high"-end one from a specialty shop.
- Boning knife: Has a thin flexible blade. I love this one for filleting and skinning fish or boning a chicken. It moves easily around the bones so I can maximize the amount of meat without it looking like it was hacked from the carcass. Some cooks prefer a long blade, but I find I have better control with a shorter blade.
- Sharpening steel: The steel will realign the fines of the edge of the blade between uses. Over time the blade will dull as it loses its edge and need sharpening. I find, as a home cook, once-a-year sharpening is all I need for an expensive chef's knife. If you've invested handsomely in these knives, enlist a professional to sharpen them.

PANS

My only "must-have" pan is a small curved nonstick pan with a heavy metal bottom. This is my go-to pan for scrambled eggs, omelettes, and over-easy eggs. The food moves around the pan with a quick hand motion and slips easily onto a plate. Invest in this kind of pan—it's game changing.

All the other pans are comparable; some tout holding heat better, and some distribute heat better. While much of this may be true, I would not run out and replace what you have already been using as it will not make a material difference to your food. If your nonstick pans are etched or scratched, it is worth replacing them. Still, there is no need to break the bank.

THERMOMETER

Yes, this is a must-have item! A thermometer tells you when food reaches a minimum temperature, ensuring it is safe to eat. Temperature is an indicator of doneness for so many foods. See the chart below for the minimum temperatures to ensure your food is safe as well as done. So often we cook to a predetermined time regardless of temperature or size of the meat and, in the case of the chicken I grew up eating, the poor chicken was unidentifiable, and the meat was dry and inedible. Just be sure the thermometer works and is accurate. Check the thermometer's accuracy by calibrating it with an ice slurry. See Appendix A on how to calibrate your thermometer.

COOKING TEMPERATURES

Whole Poultry Duck, Chicken, Turkey 82°C (180°F)	Poultry Cuts Bone-in, Boneless 74°C (165°F)	Beef, Pork, Lamb Bone-in, Boneless 71°C (160°F)	Ground Meat Pork, Beef, Poultry, Veal, Wild Game 74°C (165°F)
Steak Rare 60°C (140°F) Medium Rare 63°C (145°F)	Medium 66°C (150°F) Medium Well Done 71°C (160°F) Well Done 74°C+(165°F+)	**Wild Game** Bone-in, Boneless 74°C (165°F) Moose, Rabbit, Caribou, Elk, Reindeer, Deer	**Wild Game Birds** Whole 82°C (180°F) Ptarmigan, Goose, Partridge, Duck, Pheasant, Turkey
Fish Round Gutted, Fillets, Steaks 71°C (160°F)	**Shellfish** Scallops, Shrimp, Crab, Lobster, Clams, Mussels 74°C (165°F)	**Eggs** Whole Eggs, Egg Dishes 74°C (165°F)	

CHILLING AND COOLING

Foodborne illness can occur if food dwells within the temperature danger zone for a long period of time. The danger zone is between 4°C (40°F) and 60°C (140°F). It is the temperature where pathogens are quick to reproduce. Pathogens can cause severe illness, even death. Hot food does have to go through the danger zone to chill completely. Chilling food quickly and following safe chilling practices mitigates proliferation of harmful bacteria. The rate of chilling should be from 60°C (140°F) to 20°C (68°F) within two hours maximum and from 20°C (68°F) to 4°C (40°F) within four hours maximum.

SCALE

A small kitchen scale gives me confidence that my food will turn out. I like a scale with discrete measurements with grams shown to one decimal place for fine measurement of spices and ingredients, where a little has a big and important effect (e.g., baking soda, baking powder, salt).

BLENDER/FOOD PROCESSOR

You can see in the picture on page 9 that my blender is somewhat of a relic, but it functions to reduce large pieces to small ones, small to fine particles, and fine to purees. It works for hot or cold and dry or wet foods. My story on the blender in the picture is I bought it a garage sale in 1986, and it has outlived many upgrades. I do have a food processor as well, so it is your choice. But at least have one.

STRAINERS AND COLANDER

I am always impressed with a smooth, silky, glossy sauce accompanying any protein. The secret is to push the sauce through a fine mesh strainer. So I swear by having a few of these on hand, but in particular, look for at least one that is very fine mesh, you will quickly see how different gravy will be, easily complementing your Sunday dinner. As well, I keep a small strainer to use for straining seeds from squeezed lemons. On the other hand, a large strainer can strain water from cooked pumpkin, potatoes, and beans.

DOUBLE BOILER

This is an accessory most people do not have. I have a small one with a rounded bottom that allows me to whisk hollandaise without accumulation of the sauce in corners or on the sides. Same goes for melting chocolate and being able to control temperature. This pot makes my "must-have" list.

OTHER IMPORTANT UTENSILS

Rosle twelve-inch tongs are pure indulgence and the best tongs I have ever used. The pincers are more precise and allow for better control.

I like whisks that have medium-size wires that have some movement. I like to have a few large ones for large quantities and a few small ones to whisk a cup of mayonnaise or vinaigrette.

Look for a measuring cup set that has deep cups and no lips or spouts. Spouts don't work that well to reduce spillage and make it hard to scoop and sweep precisely.

Timers will keep you on track. I have a few small ones, but mostly I just use my watch.

A carrot peeler is good for peeling carrots and celery and uniform potatoes. A peeler with an inch-and-a-half blade can also create thin slices. Very useful!

You can buy various-sized wooden spoons and rubber spatulas at the dollar store, grocery store, or hardware store. They are so well used by almost everyone. I have one for every pot, pan, and bowl. They are invaluable for incorporating, stirring, and scraping.

9

MEASUREMENT, PORTIONS, AND COOKING MATH

The idea of measuring is eloquently simple yet when ignored is sure to thwart success. From my days as a Six Sigma Black Belt at Maple Leaf Foods, I have learned and experienced the effect consistency can have on understanding where things go wrong or right. To determine what goes right or wrong, measuring—and measuring consistently—plays a material role, no matter what is being measured and for what purpose.

For me, every food ingredient has a function, no matter what I'm making. The ingredient may be used to enhance flavour, or add texture, or create colour and appeal. The ingredient may be considered to increase nutrition or add richness. So, to ensure the purpose of each ingredient and its expectation is realized, measuring carefully and consistently each time is critical. Measuring accurately and consistently ensures that ratios of ingredients can function to improve texture, flavour, appearance, and how much you make (yield).

The secret to accurate and consistent cooking is doing it the same way time and time again. Knowing how to measure accurately, and doing so every time you make the food, is the key to consistently making the dish.

USING A SCALE
They say cooks estimate but bakers weigh. That's because the ratios of baking are more critical than in cooking. I advocate weighing everything if you can, because weighing ingredients is more efficient and more accurate. This is especially true in baking, where miniscule ingredient inaccuracies have a major effect on the outcome of any baked good. Scaling ensures consistency every time, regardless of who is doing the work!

Did you know food preparation is ultimately based on weight ratio, yet the measuring cups we use are generally based on volume? It is easy to get off track using only measuring cups, because they don't always end up being the same from one scoop to another. Weighing food gives you more precision, and since different ingredients have different densities (i.e., weight per volume), weighing food helps ensure all ingredients are treated the same. Weight is precise, so you can be confident your dish is going to turn out the same way every time.

Here is an example of how weighing ingredients helps ensure consistency at Java Jack's Restaurant & Gallery. Initially when staff made coffee, they used cups to fill the machine with coffee grounds. Every pot got a slightly different amount of coffee. This meant each pot of coffee had a slightly different flavour. Instead I trained the staff to weigh the coffee grounds (75 g per pot of coffee), and now every pot of coffee served at Java Jack's has the same consistent flavour! Guests notice the difference and appreciate the consistency of taste whenever they ask for a cup.

In this book, The Bakery chapter has both volume and weight measurements while the rest of the recipes use volume, primarily. You can try out recipes with both measurements to test the use of a scale and reconvert your own recipes to weights.

HOW TO USE A SCALE
Step 1.   Turn the scale on, and allow the zero to appear.
Step 2.   Place an empty container on the scale. Press tare (ie., subtract the weight of the container from the total weight. The scale will read zero.)
Step 3.   Place the ingredient in the container to the amount required.
Step 4.   Transfer as needed. Return to Step 1 and repeat for the next ingredient.

Rule of thumb: Clean the scale pan, and let it dry completely before using it. The weight of the ingredient or container changes as the moisture evaporates.

USING VOLUMETRIC MEASURING CUPS

Traditionally we all use measuring cups and spoons of various sizes that have been gifted or inherited. Some are fun shapes and are made of metal, glass, or plastic, and they stand the test of time. These cups are volumetric measures. I like my Tupperware measuring cups. They are a regular round shape and do not have pouring spouts. I can ensure my measurements are precise because of this detail. I have a large glass 4-cup measure as well, but I know it offers less precision. My favourite measuring spoons are oval with a deep hold. I find shallow, wide spoons are not as easy to use, and the measurements are less accurate. A kitchen teaspoon is not a teaspoon—not all spoons are made equal.

Let's go through an example of a scoop-and-sweep method to ensure better precision measuring a cup of flour:

Step 1: Using a clean, dry, 1-cup measure, scoop up enough flour to overfill the cup.
Step 2: Tap the side of the cup lightly a few times to compact the contents.
Step 3: Add more flour if necessary to fill the cup completely.
Step 4: Place the back of a butter knife with a straight edge on the edge of the cup, and sweep the knife across the top of the cup to ensure surface of the flour is even.
Step 5: Transfer the flour completely as needed. Tap the cup to transfer any residual powder or material.

USING A THERMOMETER

I can't stress enough how critical measuring temperature is for cooking safe and juicy meat or poultry. Similarly, baking doneness can best be measured by taking a temperature, though it is not the focus here. A working thermometer used in cooking meat ensures your guests will not get sick from undercooked food. As well, the thermometer will tell you if your oven is working optimally. A thermometer is an invaluable tool, and if you don't have one, drop everything, run out, and get one. Nothing fancy is needed. Order it online or get it from a local grocery or hardware store. Batteries are available from a dollar store or pharmacy and are very inexpensive.

Here's how to use a thermometer to measure temperature accurately:

Step 1: Turn on the thermometer.
Step 2: Determine the target temperature needed in degrees Fahrenheit or Celsius.
Step 3: Insert the probe into the middle and thickest part of the food. Ensure you insert 2.5 cm (1") of the probe. Wait 20 seconds for the temperature to transfer from the hot meat to the cold probe and register the temperature. (Stir liquids first so the heat is entirely distributed.)
Step 4: Remove and clean the probe immediately with soap and water. Dry completely before reusing or storing.

> Food surfaces allow heat to escape quickly. If you measure the temperature from the surface of the food, this reading does not represent the entirety of the food's temperature.

Rule of thumb: *To determine if a thermometer is working well, first change the battery and then check its calibration for accuracy. Ice water is 0°C (32°F), so make a slush of ice and water and insert the probe. Wait for 20 seconds. See Appendix A for more details.*

Convert Celsius to Fahrenheit
(°C multiply by 1.8) then add 32 = °F

Convert Fahrenheit to Celsius
(°F subtract 32) then divide by 1.8 = °C

USING A TIMER

Consider that every cooking appliance in the kitchen is outfitted with a timer. Geesh, even my watch and my phone have timers! This is no accident. Time is a major factor in ensuring delicious-tasting, exquisite-looking, safe food. Temperature, together with time (and patience), are the two most influential factors in your cooking. I have said for years, no matter if I'm making food for the masses or one plate at a time, remember time and temperature, time and temperature. Watching the clock is fine, but precision and success each time requires you to set and keep to the timer.

Let me give you an example. During a visit to France, we registered for a cooking class at Plum Lyon. We asked pastry chef and owner, Lucy, for tarte tatin, a traditional French apple pie served for dessert. She selected and weighed all the ingredients and followed her directions precisely, rolling out the dough, creating caramel, cutting the apples, and assembling the tarte tatin. After the oven preheated, she put the pie in, closed the oven, and set the timer for 40 minutes. When the timer rang, she promptly took the apple pie out of the oven without needing to check on it or open the oven during the bake. She put it in, set the timer, and left it, confident the tarte tatin would be perfect after the set time. And it was.

> *Success is a science; if you have the conditions, you get the result.*
> *-Oscar Wilde.*

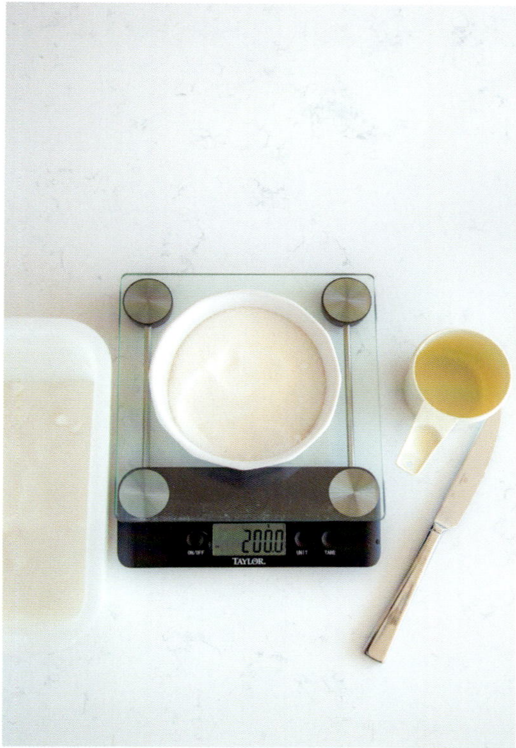
200.0
TAYLOR

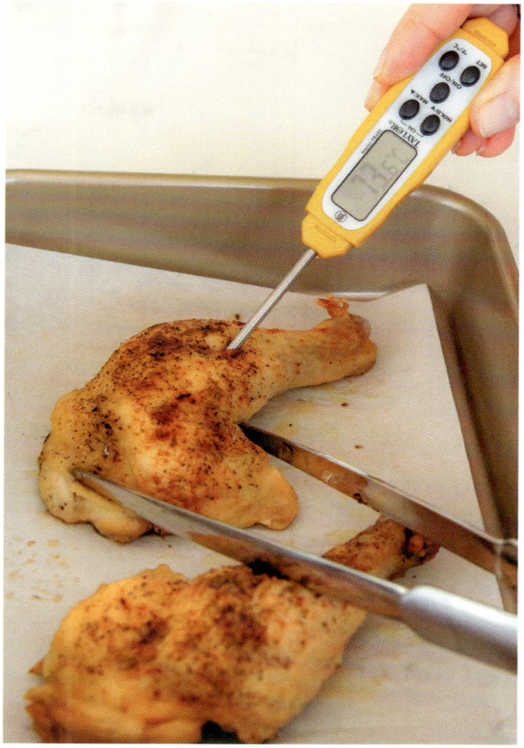

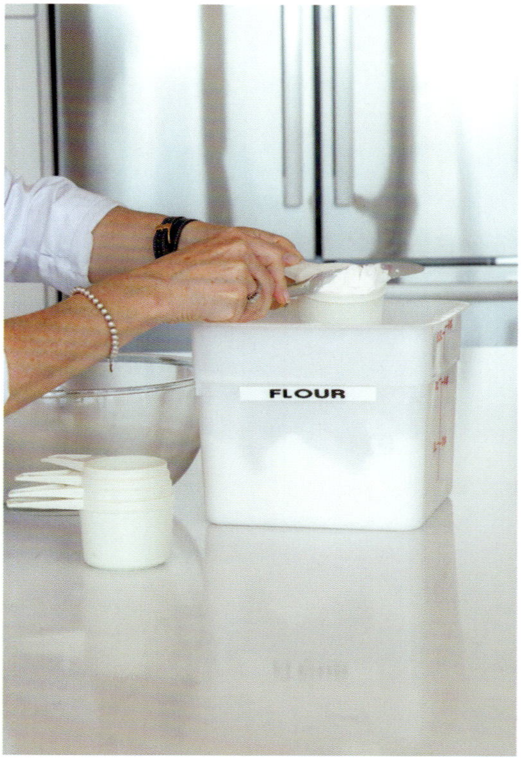
FLOUR

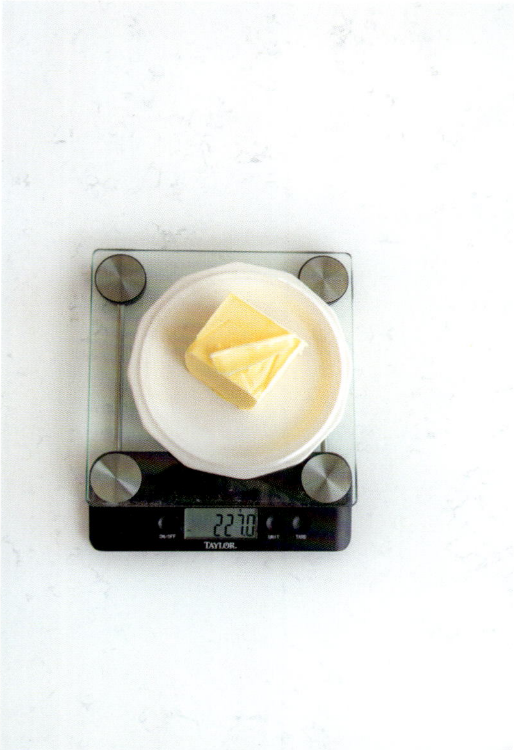
227.0
TAYLOR

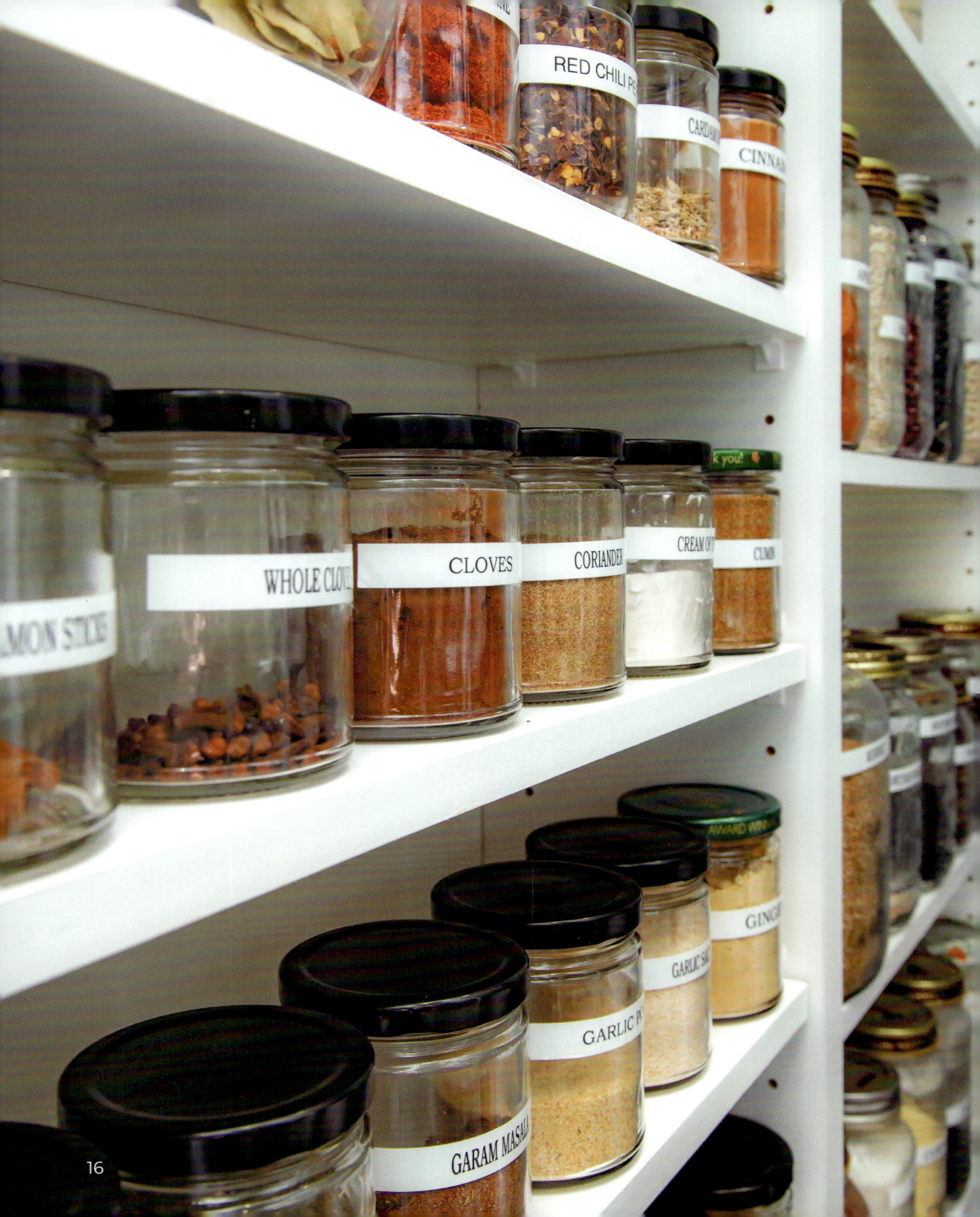
RED CHILI
CLOVES
CORIANDER
CUMIN
GARAM MASALA

MISE EN PLACE

Mise en place is a French term that means "put in place" or "gather." It refers to the preparation and setup required before cooking—the steps of getting ready and getting organized. For me it is not just about the recipe and ingredients and getting them prepared but also about a state of mind of success. Having all your ingredients measured, cut, peeled, sliced, grated, etc. before you start cooking organizes your mind as well as the food. The recipe is set out. Discard buckets are set out. Pans are prepared. Mixing bowls, tools, and equipment are in place of use. It's a technique most chefs use to assemble meals quickly and effortlessly.

When everything has a place and everything is in its place, prepare all the ingredients and organize your cooking space and what you will do with the empty containers, dirty dishes, messy hands, and garbage. Think about the cooking utensils: some are for cooking, stirring, whisking, and flipping; some are for handling raw; and some are for tasting. Think about the process, how long it will take, what will be used, what will be dirty. It's such an important concept that is often understated.

1. Get the recipe.
2. Review all the ingredients and required quantities.
3. Create a grocery list and determine what needs to be purchased and how much.
4. Determine the time required for preparation.
5. Gather ingredients.
6. Gather utensils and tools.
7. Review the procedure and how to prepare the ingredients.
8. Prepare and measure ingredients.
9. Begin cooking. Heat (or chill, if applicable) as required, ensuring proper heating and cooling times and temperatures. (See section on temperatures in Chapter 2, p. 7.)
10. Assemble ingredients and place them on a serving dish.
11. Serve, keeping hot food hot and cold food cold.
12. Store leftovers at temperatures equal to or less than 4°C. Be sure not to leave food out on the counter to cool. (See section on temperatures in Chapter 2, p. 7.)

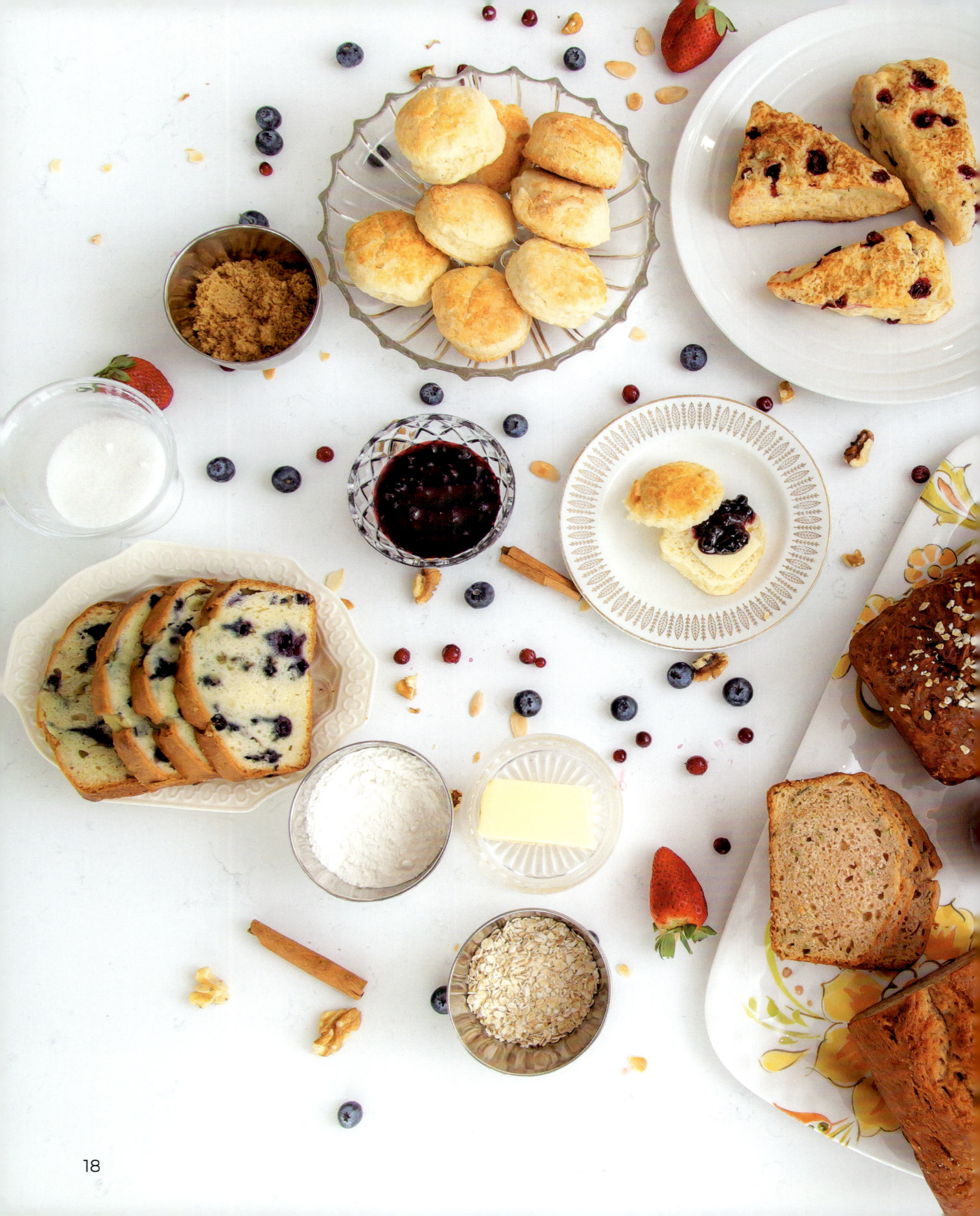

3. The Bakery

MY STORY: COMPANY'S COMING

In the 1950s it was unusual for a wife and mother to work outside the home, but my grandmother, Ivy Penney Murphy, worked as a tailor to help the family. Times were tough, and my grandparents lived modestly. "Mom," as she was known, was a talented lady in many ways, and she made a dollar go a long way. So, when company would arrive, she would quickly mix up a batch of tea buns, toss them in the oven, see them rise into flakey goodness, and within 20 minutes her guests sat into the table (an Irish way of saying they gathered around the table) for tea, homemade jam, and the warm biscuit tea buns. My mother took up this practice, and eventually so did I. I serve Mom's tea buns in a pinch at my table when company arrives and I'm in need of a quick snack that impresses. I do the same as my grandmother did and smile each time, thinking how clever she was.

THE MAGIC OF SCIENCE: LEAVENING AND BROWNING

Leavening magic: I know people often wonder what causes baked goods to be so flakey, or airy, or to rise. Ingredients like yeast, baking powder, baking soda, and eggs create the ability for a mixture to rise.

For many baked goods, where the cooking time is very fast, you need fast, active rising. For this you can choose either baking powder or baking soda, as each contains sodium bicarbonate. When it is added to a liquid such as water or milk, the sodium bicarbonate is ready to react. Baking soda needs an acidic ingredient to react quickly to and release carbon dioxide. The acidic ingredients most often used to react with baking soda are buttermilk (or milk soured with vinegar or lemon juice), yogurt, sour cream, honey, molasses, and puréed fruit (e.g., bananas and apple). But since not every recipe includes an acidic ingredient, an alternate leavening agent is needed: baking powder.

Baking powder is simply baking soda blended with just the right amount of acid in the form of a dry compound. When mixed with liquids, the baking powder dissolves allowing the baking soda and acid to react and produce carbon dioxide. To keep the two active ingredients, dry and separate, baking powder also contains cornstarch.

Recipes often include both baking soda and baking powder. This is because cakes, muffins, tea buns, and quick breads require a certain ratio of leavening agent to flour to produce the right amount of carbon dioxide, establish a matrix, and rise properly. If the recipe doesn't contain enough acid to react with the baking soda, baking powder is used to make up the difference.

Browning magic: Browning is a complex event. Even an excess amount of bicarbonate can contribute to the browning process. Caramelizing sugar, colour development of meats, and colour from enzyme reactions have different chemical reactions. Most browning, particularly for baked goods, develops because of a Maillard reaction. It takes place when ingredients that contain proteins, such as gluten from wheat flour, milk proteins, or egg proteins, and simple sugars such as fruit sugars (fructose), cane sugar (sucrose), and milk sugars (lactose), react in the presence of heat and a little water. The mixture starts out colourless, and as the heat advances and moisture evaporates, the Maillard reaction occurs, and the golden-brown colour starts to develop. In the final stages of the reaction, the food goes from red brown to very dark brown. As well, the distinctive baked aroma and flavour develop. Yum!

BISCUITS AND DUMPLINGS

Biscuit dough is so versatile. You can sweeten it for an afternoon tea snack and use it as the medium for nuts, fruit, berries, and chocolate to suit everyone's preference. Or you can spoon the dough on top of fruit compote and bake, as in a cobbler, to create a decadent, healthful dessert. You can repurpose the dough with just a little or no sugar and combine with savory ingredients such as cheese and herbs and bake, or just serve plain biscuits with butter to accompany a stew. Spoon the unsweetened savory version on a thick soup or stew and cook with the cover off or on. Rolled out, you can use biscuit dough to create an old-fashioned country pie in an instant.

- Grandmother Ivy's Tea Buns
- Dumplings for Stew
- Scones
- Strawberry Cobbler

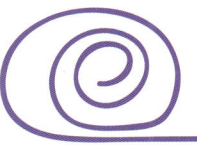

Rule of thumb: Be sure to use cold butter and handle it as little as possible, or not at all, so the fat doesn't melt in your hands before the dough matrix has a chance to form and encapsulate the little fat particles that create a flaky biscuit or dumpling.

Rule of thumb: Assume 1 to 2 tsp of baking powder or ¼ tsp of baking soda (used in combination with an acidic ingredient) for each 1 cup of flour.

Also, ½ tsp of baking soda used in combination with 1 cup of buttermilk or soured milk produces the same effect as 2 tsp of baking powder.

GRANDMOTHER IVY'S TEA BUNS

MAKES ABOUT 12 TEA BISCUITS

3 cups (450 g) flour, sifted
3 tsp (15 g) baking powder
½ tsp (2 g) salt (if butter is not salted)

1 cup (225 g) butter or margarine
1 cup (250 g) milk
1 egg, well beaten

1. Preheat the oven to 350°F.
2. Combine the dry ingredients until mixed.
3. Rub/cut butter into the dry ingredients until you have a pebbly texture.
4. Slightly beat the egg and add it to the milk.
5. Add these wet ingredients to the dry mixture and mix only until combined.
6. Turn the dough out onto a floured board or counter. Press it lightly and cut with a cutter. Don't overmix or over handle the mixture, as this will reduce the lightness of the buns.
7. Brush a little egg wash or milk or melted butter on the biscuits. Bake for 25 minutes.

Rule of thumb: Ensure the baking powder is not expired. Its ingredients do lose potency over time, especially if they're exposed to high humidity and moisture, and the buns will not rise as well.

To check the baking powder effectiveness, stir a spoonful into approximately ½ cup (125 ml) hot water. If the mixture bubbles, the baking powder is still active.

Baking powder contains sodium bicarbonate and monocalcium phosphate and reacts in the presence of moisture from milk and egg. Use baking powder for foods that require a bake time of less than 20 minutes.

Out of baking powder? Make some in a pinch.
Mix ½ tsp cream of tartar with ¼ tsp baking soda.
This replaces 1 tsp baking powder

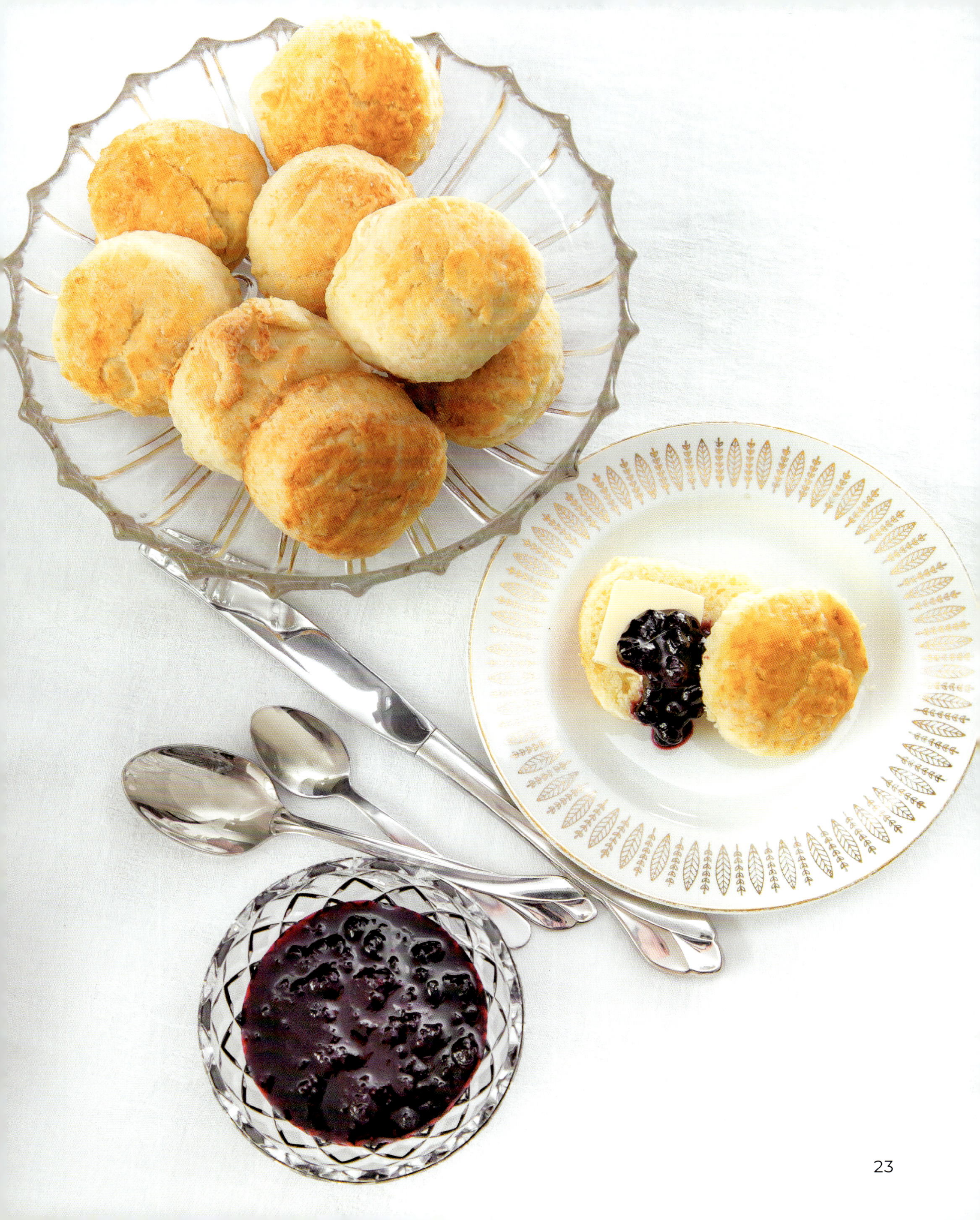

DUMPLINGS FOR STEW

MAKES 4 SERVINGS

1 ½ cups (275 g) flour
2 tsp (10 g) baking powder
½ cup (125 g) butter or margarine
½ tsp (2 g) salt
¾ cup (175 g) water

1. Mix the flour, baking powder, and salt together.
2. Rub in the butter so it looks like crumbs.
3. Add the water and mix. The dough should be wet and viscous.

Dollop the dough on top of stew or a hearty soup, one that is almost ready to serve. Maintain a continual simmer, cover, and steam cook for about 10 minutes.

Dollop the dough on top of cooked braised meat. Cook with the cover off at 350°F for 20 minutes for a little crusty topper.

25

SCONES

MAKES ABOUT 12 TEA SCONES

3 cups (450 g) all-purpose flour
¼ cup (50 g) sugar
4 tsp (12 g) baking powder
½ tsp (2.5 g) salt
2 tsp (10 mL) finely grated lemon zest
¾ cup (175 g) cold butter, cut into pieces
¾ cup (175 mL) cold milk (1% or 2%)

1 large egg
1 large egg yolk (reserve the white for brushing onto the scones)
¾ cup inclusions such as berries, chopped apple, chocolate chips or coconut or any others you like

1. Preheat the oven to 375°F (190°C) and line one or two baking trays with parchment paper.
2. Sift the flour, sugar, baking powder, and salt into a large mixing bowl. Add the grated lemon zest and mix in. Rub in the butter until no large pieces are visible and the mixture is a rough, crumbly texture.
3. In a separate bowl, whisk the milk, egg, and egg yolk together. Make a well in the centre of the flour mixture and add the milk and egg mixture all at once. Mix everything together just until it starts to come together but is still rough and crumbly.
4. Turn the dough out onto a clean surface and sprinkle the inclusions. Use your hands to bring the dough together and flatten and fold the dough over on itself a few times.
5. Flatten the dough with your hands, to just under 1" (2.5 cm) thickness, and use a 2 ½-inch (7.5 cm) round cutter to cut out the individual scones. Arrange the scones on the baking tray(s), leaving 2 inches (5 cm) between each scone. Brush the tops of the scones with the egg white and sprinkle with cinnamon sugar.
6. Bake the scones for about 18 minutes, until golden brown on the bottom and lightly browned on top.

STRAWBERRY COBBLER

MAKES 6–8 SERVINGS

6 cups (1200 g) fresh strawberries, hulled, cut in half
½ cup (100 g) sugar
2 tbsp (30 g) cornstarch
½ cup (100 g) sugar
½ cup (125 g) soft butter or margarine

2 large eggs
1 tsp (5 g) vanilla
1½ cup (125 g) flour
1 tsp (5 g) baking powder
½ cup (125 g) sliced almonds

1. Preheat the oven to 350°F. Lightly grease a 9x11 glass baking dish with cooking spray oil.
2. Toss berries in the ½ cup sugar and cornstarch and add to the baking dish.
3. Mix the rest of the ingredients together and combine well.
4. One spoonful at a time, spread the topping over the strawberries.
5. Bake for 25–30 minutes until the topper is golden and mixture is bubbling.

LOADS OF LOAVES

The concept of quick breads is about how fast the preparation is and how quickly the bread rises compared to traditional yeast bread. Preparation is just 10 minutes when you have the mise en place. The magic is in the combination of the leavening and the work that happens in the oven during the cook time of almost an hour. The leavening is a little bit of quick leavening reaction at the beginning of the cook and a slower reaction to hold it all the way though cooking the large-diameter loaf. Here is the base recipe for almost all loaves, sweet or savory. So have some fun trying out different flavours!

- Classic Quick Bread Loaf
- Freestyle Loaves
  (add your own berries)
    - Zucchini Loaf
    - Cinnamon Loaf
    - Blueberry Cream Cheese Loaf
    - Banana Bread
- Irish Soda Bread

Rule of thumb: If no sugar is added to the mix, browning will not happen naturally. More often it scorches, creating acrid, bitter, burnt off-flavours.

CLASSIC QUICK BREAD LOAF

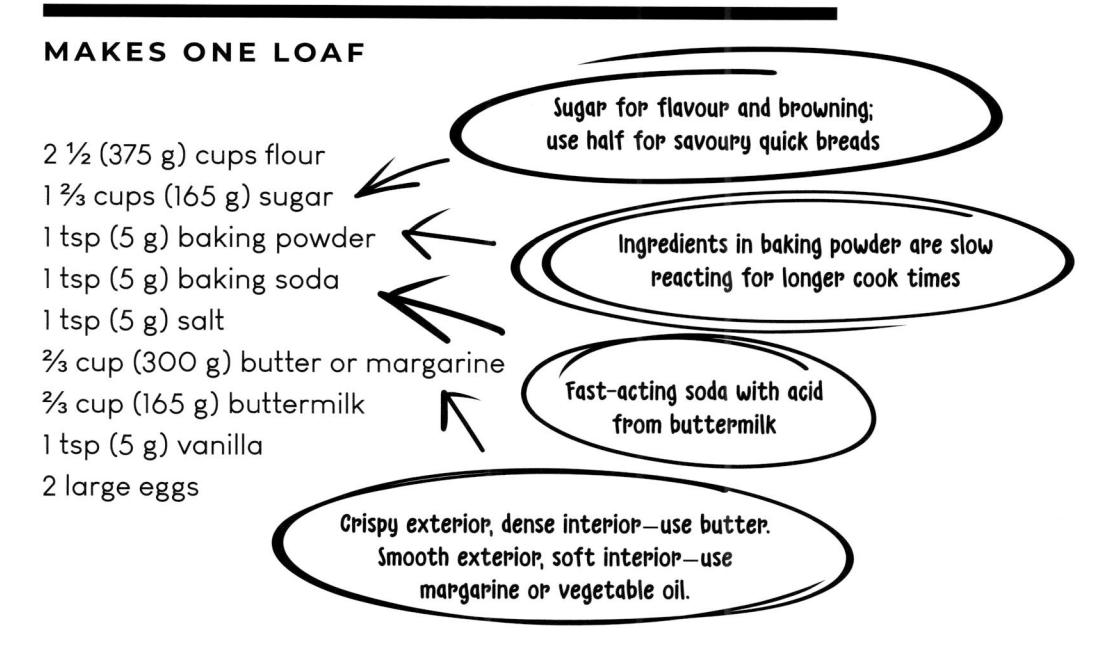

MAKES ONE LOAF

2 ½ (375 g) cups flour
1 ⅔ cups (165 g) sugar
1 tsp (5 g) baking powder
1 tsp (5 g) baking soda
1 tsp (5 g) salt
⅔ cup (300 g) butter or margarine
⅔ cup (165 g) buttermilk
1 tsp (5 g) vanilla
2 large eggs

Sugar for flavour and browning; use half for savoury quick breads

Ingredients in baking powder are slow reacting for longer cook times

Fast-acting soda with acid from buttermilk

Crispy exterior, dense interior—use butter. Smooth exterior, soft interior—use margarine or vegetable oil.

1. Preheat the oven to 350°F. Grease a loaf pan by spraying the pan with spray oil. Line the bottom with parchment paper.
2. Mix the flour, baking powder, baking soda, and salt together. In another bowl, cream the butter and sugar. Add the eggs and then vanilla Beat for 2 minutes until fluffy.
3. Add half the flour mixture and all the buttermilk to the butter-sugar-egg mixture. Repeat, ending with the final third of the flour mixture.
4. Add the spices and other ingredients you're including.
5. Bake for 50 to 60 minutes to 1 hour at 350°F.

How to make your own buttermilk
Take 1 cup (250 g) milk and add 1 tbsp (15 g) lemon juice or 1 tsp (5 g) vinegar. Let the mixture stand 10 minutes, and watch the milk get thick and begin to curdle. Stir vigorously before using.

FREESTYLE ZUCCHINI LOAF

ADD:
2 cups (500 g) grated zucchini
1 ½ tsp (7 g) cinnamon
1 tsp (5 g) vanilla
¼ cup (65 g) raisins

CINNAMON BREAD

ADD:
Extra ½ cup (125 g) buttermilk into the mix.
Mix 3 tbsp (45 g) brown sugar with 1 tbsp (15 g) cinnamon. Sprinkle over the top of the mix and cut into the mixture, swirling back and forth with a knife.

BLUEBERRY CREAM CHEESE

ADD:
2 cups (500 g) fresh or frozen blueberries
1 tsp (5 g) vanilla
½ cup (125 g) cream cheese
1 tbsp flour

Add the cream cheese and vanilla in with the butter when blending. Be sure the berries are dry and dust with the flour. Fold the berries into the mixture.

BANANA BREAD

ADD:
2 cups (450 g) mashed ripe bananas
1 ½ tsp (7 g) cinnamon
¾ tsp (4 g) nutmeg
½ cup (125 g) chopped walnuts

IRISH SODA BREAD

MAKES 2 LOAVES

5 cups (750 g) all-purpose flour
¾ cup (75 g) oatmeal
2 ½ tsp (12 g) baking soda
1 tsp (5 g) salt
2 ½ tbsp (38 g) brown sugar

¾ cup (175 g) molasses
¼ cup (65 g) butter
1 ¾ cup (435 g) milk
1 cup (250 g) dark stout beer, such as Guinness

1. Preheat the oven to 350°F (170°C). Generously grease the pans with cooking spray oil.
2. Add together the milk, molasses, and beer. Add the oatmeal, and set aside for 10 minutes so the oatmeal can hydrate.
3. Mix all the dry ingredients together and rub in the butter until it is the consistency of breadcrumbs.
4. Add the wet oatmeal mixture and mix until you have a wet dough. Transfer the dough to two standard loaf pans. Brush the loaves with egg wash (see p. 189) and dust with a little oatmeal.
5. Bake for 45–50 minutes at 350°F. (170°C). Brown sugar and molasses are the sugars for colour development.

Rule of thumb: No baking powder—why not? What's creating the leavening through to the end of the baking period? It's the molasses and beer.

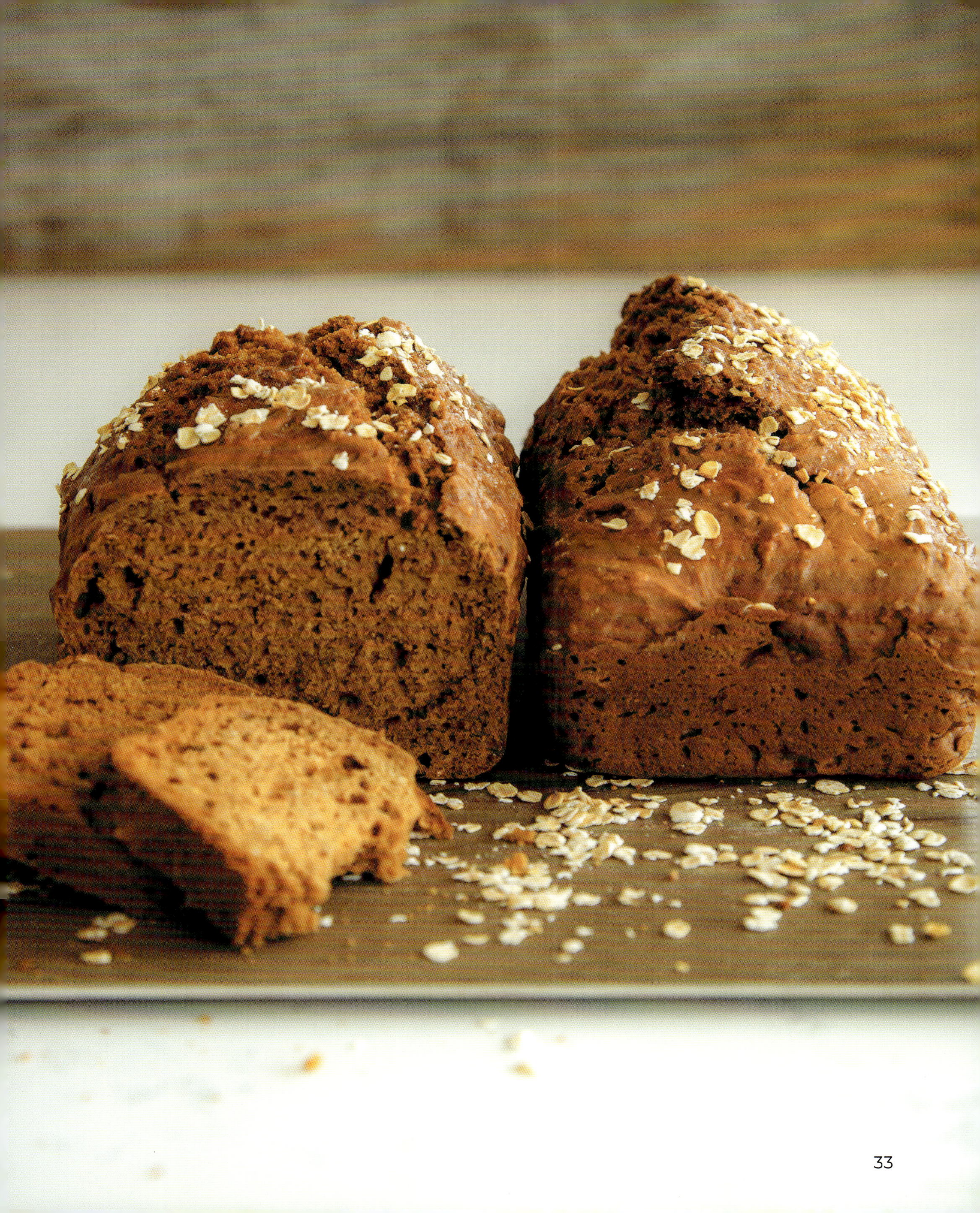

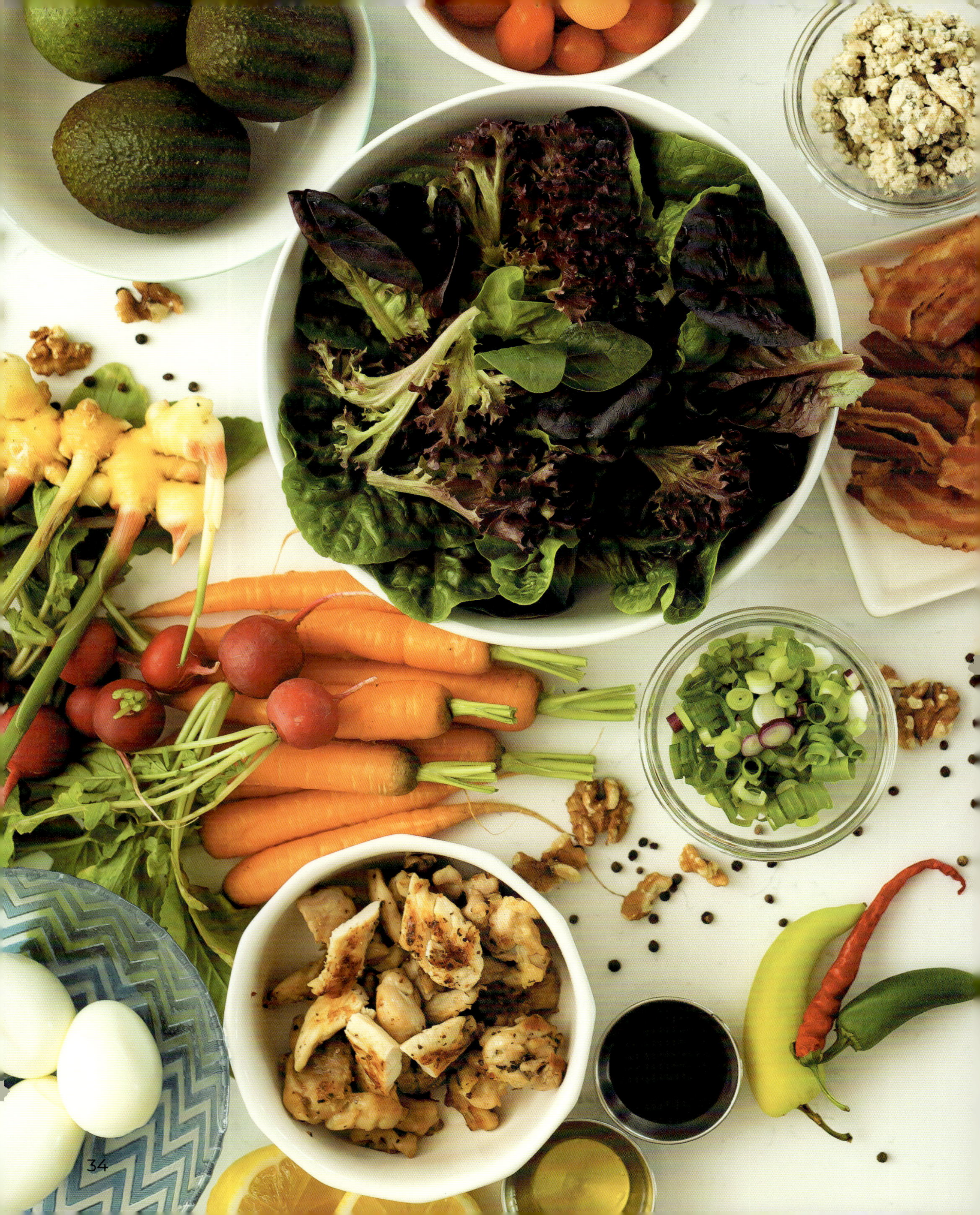

4. Building Blocks for Salads

MY STORY: EATING SALAD

Salad took over from potatoes as a staple in our household. My daughters and I assembled salads packed with natural flavours, textures, and colours. We experimented with different varieties of lettuce and greens (I do not like the texture or flavour of uncooked kale). Often, we didn't use a vinaigrette and ate salad undressed, savouring the flavour of each ingredient together and individually.

We have since added vinaigrette and dressings, and I dress our salads these days. The flavours of the ingredients just come alive when dressed in a combination of sweet, tangy, or creamy dressings. Not to mention there are many ways to mix it up, and I'm still surprised how fresh and new salad becomes.

THE MAGIC OF SCIENCE: EMULSIONS

Preparation of the vinaigrette or dressing is certainly intriguing. You can see emulsification of the oil and vinegar as it thickens in front of your eyes. It's like magic as the emulsifying components enrobe the dispersed fat droplets and hold on to the dispersed acidic component. For a vinaigrette, Dijon mustard has emulsion abilities. For mayonnaise, it's the lecithin contained in the yolk that helps with all the emulsion work. The perfect dispersion of the oil and liquid phases creates uniformity for the best flavours and aromas.

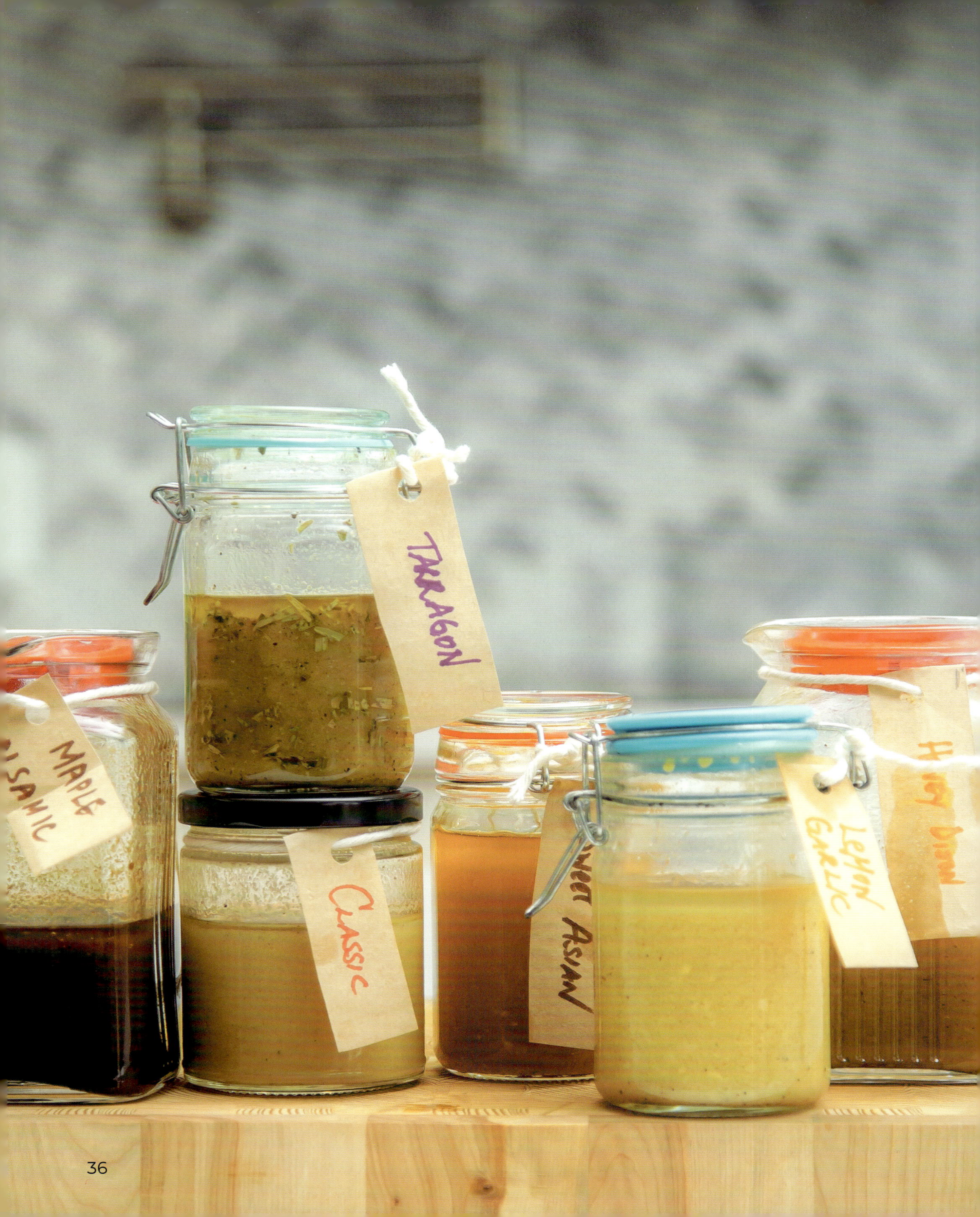
MAPLE
TARRAGON
CLASSIC
ASIAN
LEMON GARLIC

VINAIGRETTES

The classic vinaigrette is comprised of a few ingredients: oil, vinegar, Dijon mustard, salt, and pepper. Flavour variations are created by varying the flavours of the oil and vinegar. The flavours improve with a little garlic or shallot and can be changed by enhancing sweetness with honey or fruit syrups. The mustard has emulsification properties that mix oil with vinegar, so they don't separate. The classic proportions are one part vinegar to three parts oil.

- Classic Vinaigrette
- Honey Dijon Vinaigrette
- Maple Balsamic Vinaigrette
- Tarragon Vinaigrette
- Lemon Garlic Vinaigrette
- Sweet Asian Dressing

Rule of thumb: It is the homogeneity of the dressing that ensures each bite contains the right proportion of each ingredient to deliver equal and consistent flavour. But getting oil and vinegar to mix and hold together requires the addition of an emulsifier, which allows an aqueous solution such as water or vinegar to combine with oil or dairy products. Some emulsifiers are better than others. Some need agitation or mechanical action such as shaking or whisking for better dispersion. Aqueous ingredients such as vinegar, lemon juice, tomato juice, and even apple juice all have some acidity, so do not mix well with oil. An emulsifier such as mustard pulls these two kinds of ingredients together to create a delicious vinaigrette. The end result should be thick enough that it clings to the salad ingredients, and you don't find most of the dressing at the bottom of the bowl.

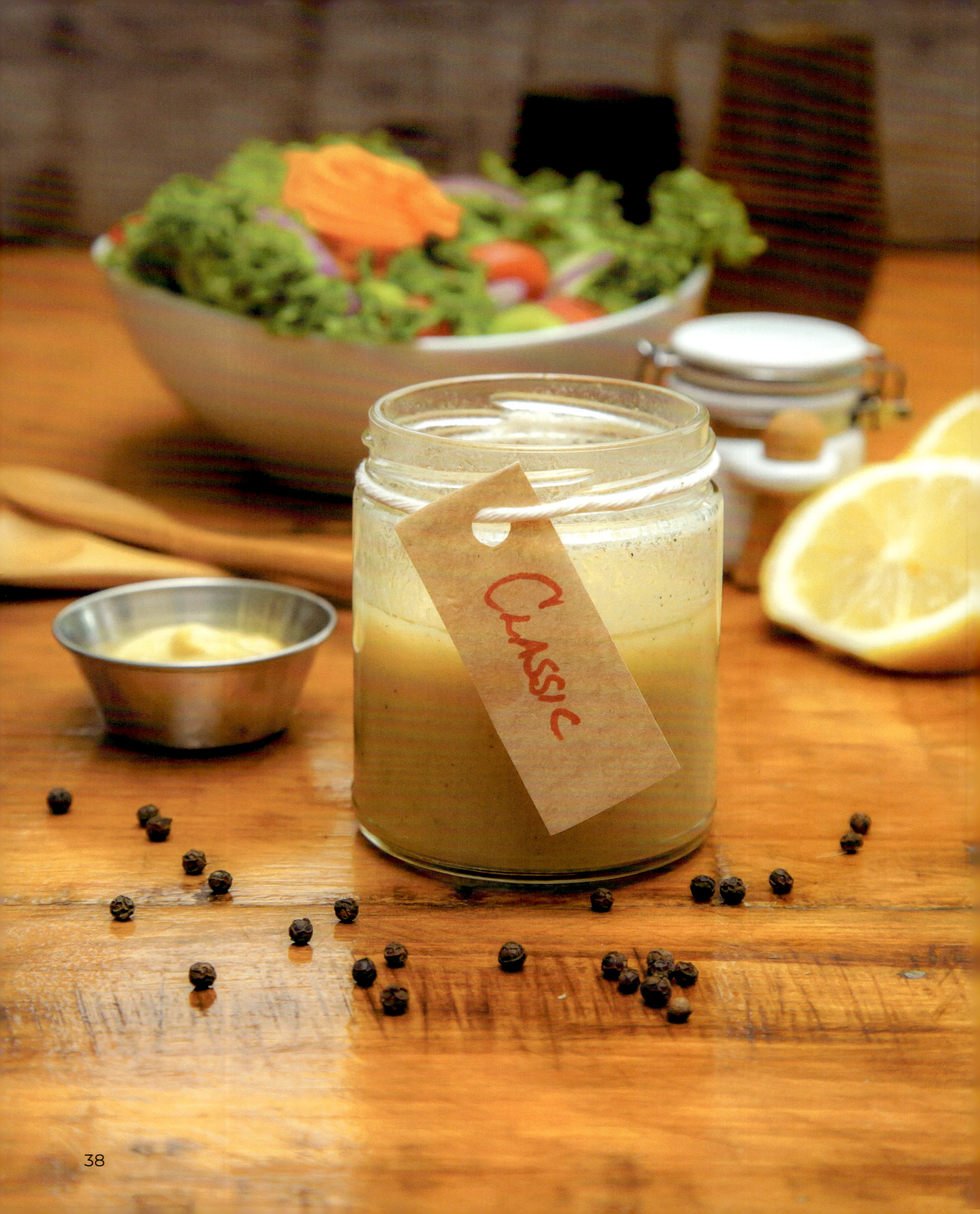
Classic

CLASSIC VINAIGRETTE

MAKES ½ CUP (4 SERVINGS)

2 tbsp vinegar (or 1 ½ tbsp lemon juice)
6 tbsp oil
2 tsp Dijon mustard
¼ tsp salt
⅛ tsp pepper

One part vinegar to two parts oil

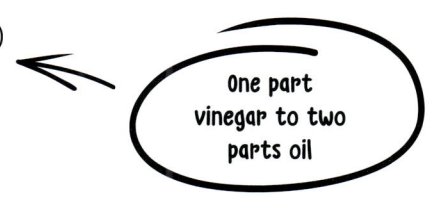

1. In a small bowl, whisk together all the ingredients except the oil.
2. Gradually whisk in the oil in a steady stream until it's blended in and the dressing has thickened.
3. Taste and adjust seasoning if necessary.

Rule of thumb: If the vinaigrette separates, check your proportions. If it's too sharp, add salt and more oil. If it's oily, add more vinegar.

Use 2 tablespoons of vinaigrette per serving of salad.

For more on classic vinaigrette, check out the "How To" video on YouTube.

HONEY DIJON VINAIGRETTE

MAKES ½ CUP (4 SERVINGS)

1 tbsp honey
1 ½ tbsp red wine vinegar
¾ tbsp Dijon mustard
½ tbsp minced shallots

¼ tsp salt
⅛ tsp ground black pepper
3 tbsp extra virgin olive oil

1. In a small bowl, whisk together the honey, vinegar, shallots, Dijon mustard, salt, and pepper.
2. Whisking constantly, slowly add the oil in a steady stream.
3. Taste and adjust seasoning if necessary.

MAPLE BALSAMIC VINAIGRETTE

MAKES ½ CUP (4 SERVINGS)

2 tbsp balsamic vinegar
1 tbsp maple syrup
1 tsp Dijon mustard

¼ tsp salt
6 tbsp extra virgin olive oil

1. In a small bowl, whisk together the vinegar, maple syrup, Dijon mustard, and salt.
2. Whisking constantly, slowly add the oil in a steady stream.
3. Taste and adjust seasoning if necessary.

TARRAGON VINAIGRETTE

MAKES ½ CUP (4 SERVINGS)

2 tbsp tarragon vinegar
2 tbsp sugar
½ tsp salt
½ tsp ground black pepper
½ tsp Dijon mustard
¼ tsp Tabasco sauce
1 tbsp fresh tarragon, chopped
4 tbsp extra virgin olive oil

1. In a small bowl, whisk together the vinegar, sugar, Dijon mustard, salt, pepper, and Tabasco sauce.
2. Whisking constantly, slowly add the oil in a steady stream. Add the tarragon and set aside.
3. Taste and adjust the seasoning if necessary.

LEMON GARLIC VINAIGRETTE

MAKES ½ CUP (4 SERVINGS)

2 tbsp fresh lemon juice
3 cloves fresh garlic, minced
½ tsp Dijon mustard
½ tsp salt
8 tbsp extra virgin olive oil

1. In a small bowl, whisk together the lemon juice, garlic, Dijon mustard, and salt.
2. Whisking constantly, slowly add the oil in a steady stream.
3. Taste and adjust seasoning if necessary.

SWEET ASIAN VINAIGRETTE

MAKES ½ CUP (4 SERVINGS)

2 tbsp rice vinegar
1 tbsp peach cantaloupe marmalade (or apricot jam)
1 tsp finely grated fresh gingerroot
1 tbsp fish sauce
1 tsp hot sauce
2 tbsp fresh mint, chopped
4 tbsp extra virgin olive oil

1. In a small bowl, whisk together the rice vinegar, marmalade, ginger, and hot and fish sauces.
2. Whisking constantly, slowly add the oil in a steady stream.
3. Add the chopped mint. Mix, taste, and adjust the seasoning if necessary.

43

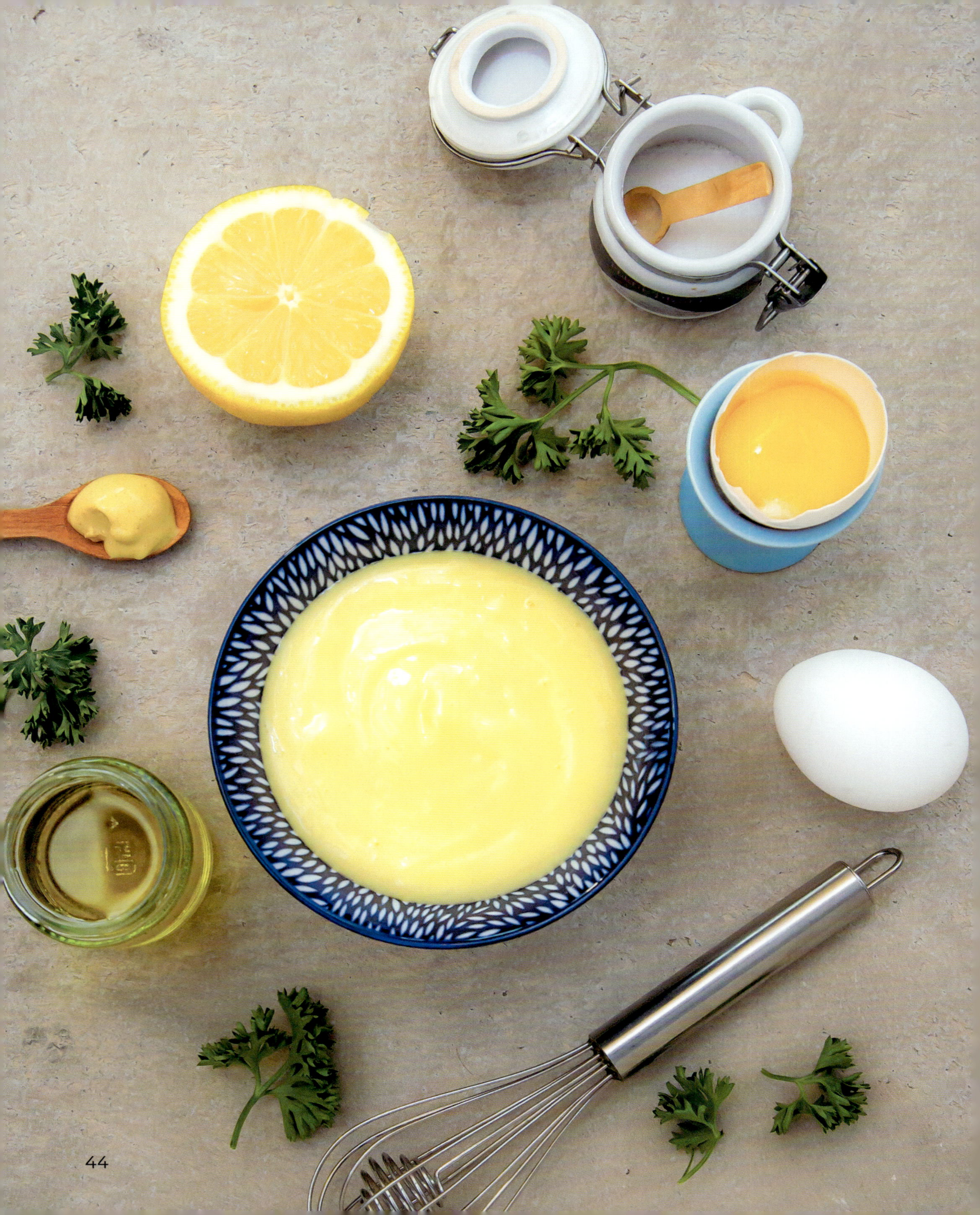

MAYONNAISE

Mayonnaise is a cold emulsified sauce made from egg yolks. It's the base for most creamy dressings and classic variations, such as aioli and Caesar, that have easy recognition and taste great. The emulsion is created when acid such as lemon juice or vinegar is whisked vigorously into egg yolks to expose the yolks' emulsifying properties and small droplets of fat are whisked and evenly dispersed into the yolk-acid mixture. Mayonnaise will hold for weeks in the refrigerator and can be used for all sorts of variations, even on sandwiches, in potato salad, green salads, and sauces. Once you make your own mayonnaise, you will never buy it again.

- Classic Mayonnaise
- Garlic Aioli
- Creamy Caesar Salad Dressing
- Creamy Avocado Dressing

Rule of thumb: *The success of mayonnaise is in exposing the lecithin in the egg yolk and dispersing it with lemon juice through vigorous whisking or blending. One large egg yolk can emulsify up to a full cup of oil without coalescing (breaking) if well prepared. The yolk colour goes from a bright to pale yellow when it is ready to accept the first droplets of fat (usually an oil).*

Those first 20 droplets or so of fat need patience and good whisking action to disperse around and completely through the yolk and acid mix. Then the mix will accept the rest of the oil in a steady stream. One tablespoon of oil can be broken up and dispersed into about 30 billion droplets with a whisk.

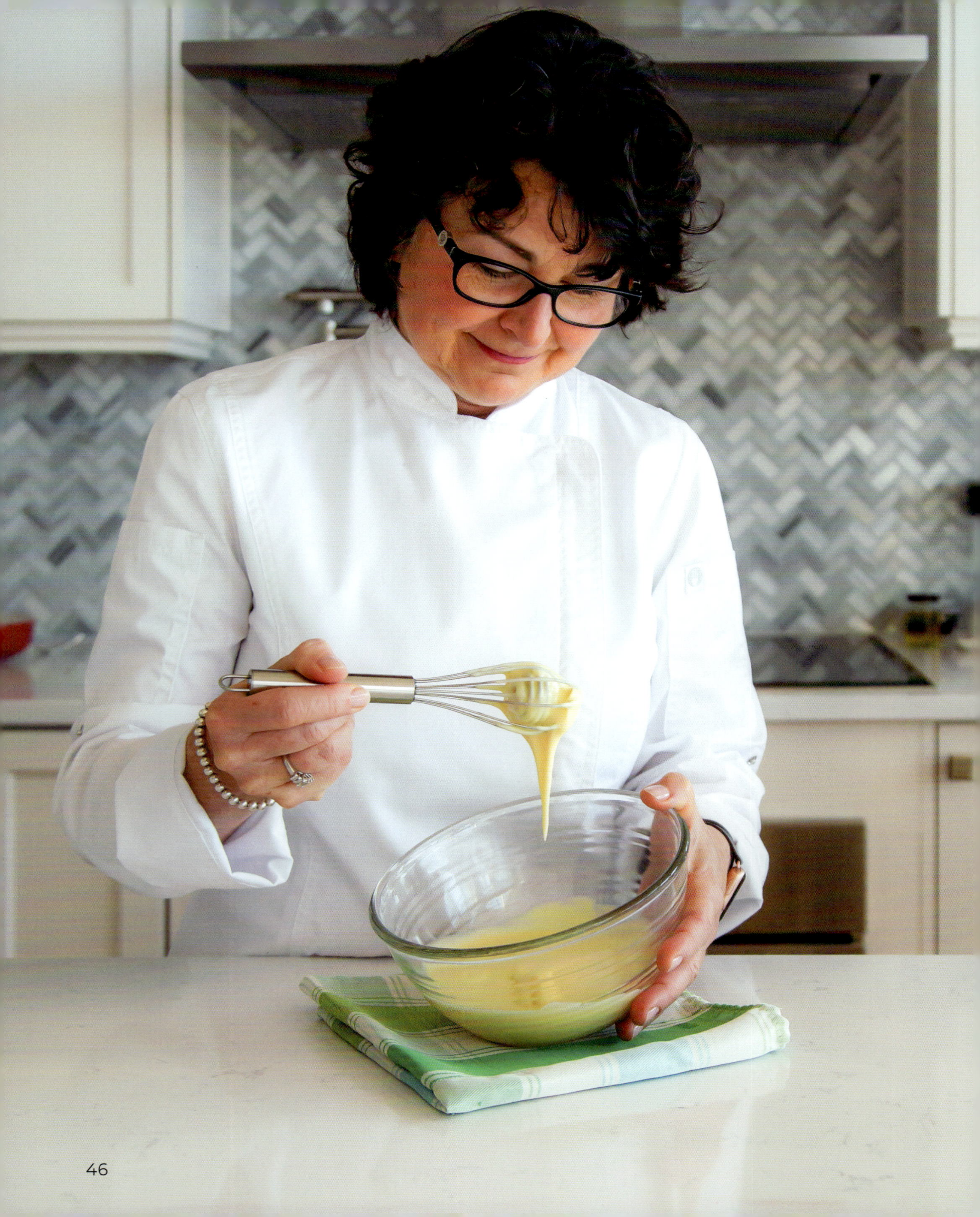

CLASSIC MAYONNAISE

MAKES 1 CUP

1 large egg yolk
¾ cup oil
2 tsp lemon juice (or white-wine vinegar)
½ tsp Dijon mustard (optional)
Dash of salt and white pepper

(1 egg yolk to ¾ cup oil)

1. In a small bowl, whisk the egg yolks together with half the lemon juice (and mustard) with a little salt and pepper. Whisk until the mixture is thick and pale yellow. It helps to set the bowl on a cloth to keep the bowl from moving as you whisk.
2. While whisking, add the oil, drop by drop. After about 20 drops or 2 tablespoons, add the rest of the oil in a slow, steady stream. Whisk vigorously and constantly.
3. Add the remaining lemon juice and season to taste.

Rule of thumb: If the mayo separates (i.e. curdles), whisk up another egg yolk with lemon and add the curdle mayonnaise mix into the egg yolk-lemon mixture one spoonful at a time. If the new mixture does not thicken, check your proportions, add a few drops of boiling water to warm it, or keep whisking.

Stored in the refrigerator, this mayo will last a month.

For more on mayonnaise and garlic aioli, check out the "How To" video on YouTube.

GARLIC AIOLI

MAKES 1 CUP

1 cup classic mayonnaise
(without the mustard)
2 cloves garlic
½ tsp sea salt

1. Peel the garlic cloves and chop finely.
2. Add the salt to the garlic on the cutting board and scrape the knife across the garlic pieces to release the juices until the garlic and salt become a paste.
3. Mix the paste into the mayonnaise. Chill and serve.

CAESAR DRESSING

MAKES ABOUT 1 ¼ CUP

1 cup classic mayonnaise
(without the mustard)
1 clove garlic, minced
1 tsp capers

2 tbsp lemon juice
½ tsp Worcestershire sauce
¼ cup grated parmesan cheese
Sea salt and cracked black pepper

1. In a food processor or blender, blend the mayo with the capers, Worcestershire sauce, and lemon juice until smooth.
2. Add in cheese and mix. Add salt and cracked black pepper to taste.
3. Taste and adjust seasoning if necessary.

CREAMY AVOCADO DRESSING

MAKES ABOUT 1 ¼ CUP

½ medium ripe avocado, peeled and pitted
1 cup classic mayonnaise
1 tbsp green onions, chopped
1 clove garlic
2 tbsp parsley
1 tbsp lemon juice
Salt and pepper to taste

1. In a food processor or blender, blend the avocado, mayonnaise, green onion, garlic, and parsley until smooth.
2. Add in lemon juice, salt, and pepper to taste.

COMPOSING A SALAD

A salad, by definition, is a mixture of small pieces of food usually seasoned with a dressing. Here, our focus is on green salad composed of lettuce, leafy and vegetables, nuts, cheese, and fruit. The basics of a good salad are really very simple: crunch, colour, height. Crunch comes from a selection of vegetables such as celery or carrots and nuts. Colour should be vibrant and complementary. Add some height and create some texture by varying the size and shape of the vegetables, such as by spiraling zucchini or cutting broccoli stalks into thin julienne sticks. Dress the lettuce and green vegetables with vinaigrette or dressing, then add other ingredients. Top specially with the ingredients that name the salad.

- Simple Green Salad with Classic Vinaigrette
- Orange and Almond Salad
- Easy Caesar Salad
- Roasted Beet and Goat Cheese Salad
- Fresh Pear and Roasted Pecan Salad
- Cobb Salad with Mustard Vinaigrette
- Bean Salad and Gourmet Tangy Bean Salad
- Potato Salad and German Potato Salad

Rule of thumb: Consider the application of the salad: side or main course? Is it for lunch or dinner? As a main dish for dinner, add some protein such as chicken, bacon, or eggs. As a side dish, have as many or as few ingredients as you like. There is ample opportunity to freestyle.

Clean and dry all the vegetables, especially the lettuce. Otherwise, the dressing and vinaigrette will not cling to the vegetables and the flavours will be diluted. Use a sharp knife to cut the ingredients. For maximum flavour impact in every bite, shred or rough chop all the pieces small enough that they can all fit on a fork at the same time, like a slaw.

SIMPLE GREEN SALAD

MAKES 4 SERVINGS

1 head green lettuce (8 cups mixed greens)
12 cherry tomatoes, halved
¼ red onion, sliced thinly
¼ cucumber, quartered and sliced
1 stalk celery, sliced thinly
½ carrot, spiralized or julienned
½ cup classic vinaigrette (see p. 39)

Colours: green, orange, red
Crunch: celery, carrots
Height: lettuce, carrots

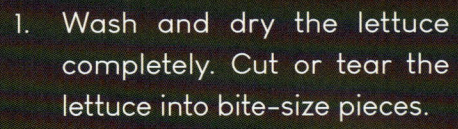

1. Wash and dry the lettuce completely. Cut or tear the lettuce into bite-size pieces.
2. Add the celery and cucumber, and toss with the vinaigrette just before serving the salad so every leaf of lettuce is coated with vinaigrette.
3. Portion and top with red onion, tomatoes, and spiralized carrots. Drizzle each salad with a little extra vinaigrette.

Rule of thumb: Dress the salad vegetables just before serving. Otherwise the acid from the vinaigrette will quickly wilt the lettuce and any delicate green vegetables.

Use two cups of lettuce or greens and two tablespoons vinaigrette per main course serving. Use half these quantities for a side salad.

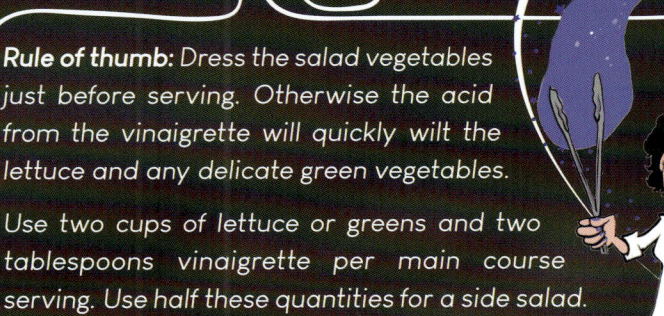

For more on composing a green salad, check out the "How To" video on YouTube.

FREESTYLE
ORANGE ALMOND SALAD

MAKES 4 SERVINGS

½ head green or red leaf lettuce
½ head Romaine lettuce
2 green onions, thinly sliced
2 tbsp fresh parsley, finely chopped
½ cup almonds, toasted

1 cup celery, thinly sliced
2 medium oranges, segmented (or 1 can mandarin oranges, drained)
½ cup Sweet Asian or Tarragon Vinaigrette (see pp. 41–42)

1. Place the lettuces, celery, green onion, and parsley in a bowl.
2. Add the vinaigrette and toss to coat.
3. Serve the dressed lettuce in bowls and top with toasted almonds and orange segments. Enjoy!

CREAMY CAESAR SALAD

MAKES 4 SERVINGS

½ cup Caesar dressing (see p. 48)
1 head Romaine lettuce, washed, dried, cut into bite-size pieces
1 cup cooked bacon, chopped
1 cup homemade croutons

½ cup shaved Parmigiano Reggiano
1 medium lemon, cut in wedges
Cracked black pepper
Place the lettuce in a bowl.

1. Place lettuce in a large bowl.
2. Add the dressing, bacon, croutons, and Parmigiano Reggiano, and toss until all ingredients are lightly coated with the dressing.
3. Serve immediately, squeeze a lemon wedge over the salad, and dust with a little cracked pepper. Enjoy!

HOW TO MAKE YOUR OWN CROUTONS
Tear bread. Toss the bread in a mixture of olive oil, salt, pepper, and herbs (thyme or rosemary) until the pieces are fully coated. Bake at 375°F for 10–15 minutes until crispy.

For more on Caesar salad, check out the "How To" video on YouTube.

FRESH PEAR AND ROASTED PECAN SALAD

MAKES 4 SERVINGS

½ cup Maple Balsamic Vinaigrette (see p. 40)
½ cup whole pecans, toasted
½ cup blue or gorgonzola cheese, crumbled
2 ripe pears, sliced

1 head green leaf lettuce (8 cups mixed greens), washed dried, cut into bite-size pieces

1. Place the lettuce in a bowl.
2. Add the vinaigrette and toss to coat.
3. Serve the dressed lettuce in individual bowls and top with toasted pecans and crumbled blue cheese. Arrange the pears on top like a fan. Drizzle a little more vinaigrette over the pears. Enjoy!

TWO WAYS TO TOAST NUTS
Oven: Preheat the oven to 400°F. Spread the nuts on a baking sheet and bake for 8–10 minutes until they are slightly browned.
Frying pan: Put a frying pan on the stove. Add nuts until they just cover the surface of the pan. Heat the pan slowly from low to medium heat. Toss the nuts occasionally until most are golden brown.

ROASTED BEET AND GOAT CHEESE SALAD

MAKES 4 SERVINGS

½ cup Honey Dijon Vinaigrette (see p. 40)
½ cup walnut pieces, toasted
½ cup goat cheese, crumbled
2 medium beets, roasted, cooled, and diced

1 head green leaf lettuce (8 cups mixed greens), washed, dried, cut into bite-sized pieces

1. Place the lettuce in a bowl.
2. Add the vinaigrette and toss to coat. Serve the dressed lettuce in individual bowls and top with walnuts, goat cheese, and beets. Enjoy!

HOW TO ROAST BEETS
Wash the beets, and boil until a fork can penetrate to the centre. Cool. Peel off the skins, dice or quarter the beets, and roast at 400°F for 8–10 minutes.

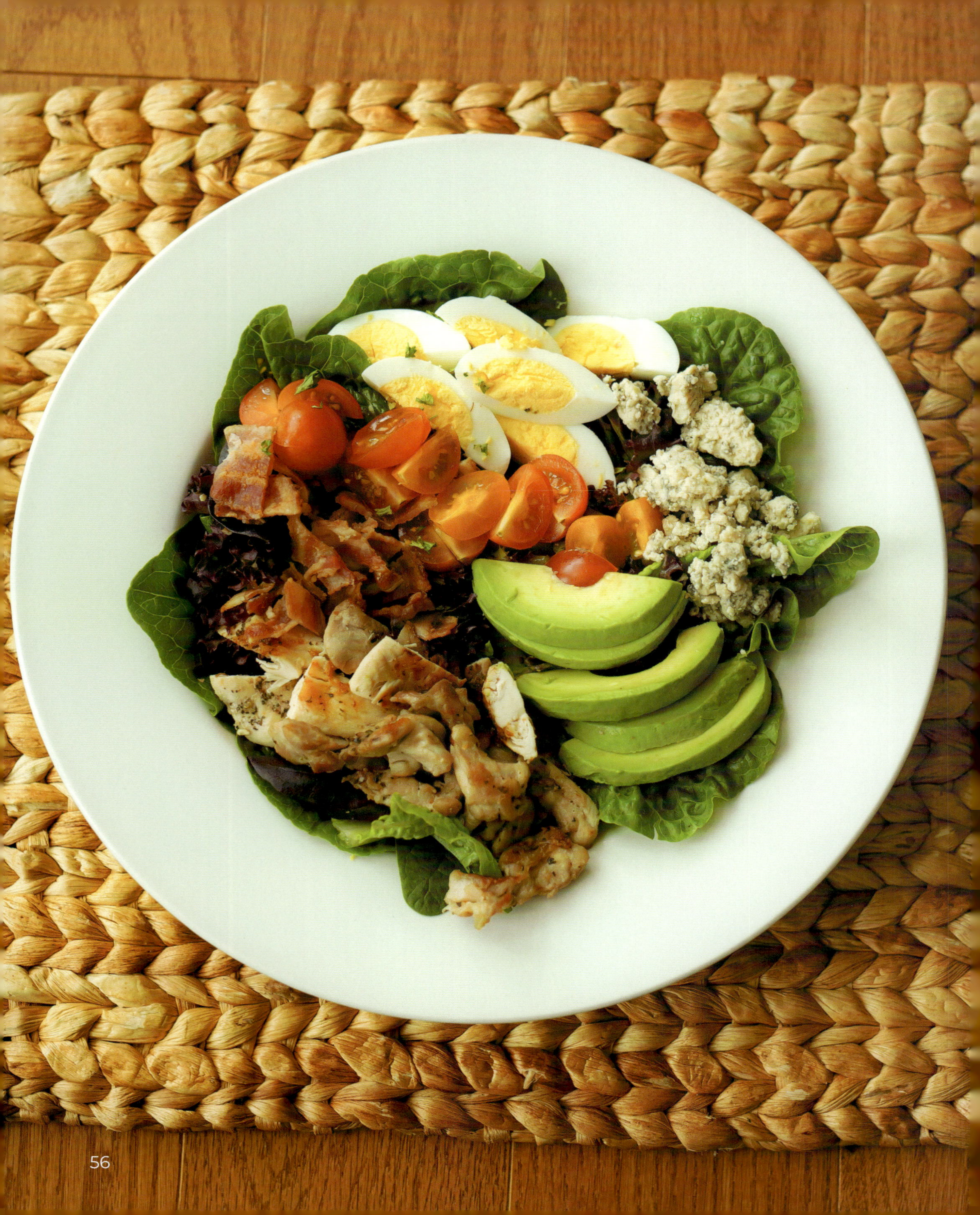

COBB SALAD

MAKES 4 SERVINGS

½ cup Mustard Vinaigrette (see below)
4 eggs, hard boiled (see p. 129), peeled and cut in quarters
8 strips bacon, cooked crisp and chopped
2 chicken breasts, grilled, chilled, and sliced or diced
1 head green leaf lettuce (or 8 cups mixed greens, watercress, and frisée), cut or torn into bite size pieces
12 cherry tomatoes, halved
1 cup blue or gorgonzola cheese, crumbled
1 medium avocado, pitted, peeled, and sliced

1. Toss the lettuce with half the vinaigrette.
2. Place the dressed lettuce in individual salad bowls or plates.
3. Assemble the other salad ingredients separately on top of the lettuce. Drizzle the rest of the vinaigrette over the finished salad, then serve immediately.

> Cut the avocado only when you're ready to serve it to avoid discolouration (caused by exposure to the oxygen in the air).

MUSTARD VINAIGRETTE

MAKES ABOUT ½ CUP

2 tbsp white-wine vinegar
½ tsp salt
1 tbsp Dijon mustard
6 tbsp extra virgin olive oil

1. In a bowl, whisk together the vinegar, salt, and mustard.
2. Whisking constantly, slowly add the oil in a stream.
3. Taste and adjust the seasoning if necessary.

If using a jar, add all the ingredients and shake vigorously until mixed.

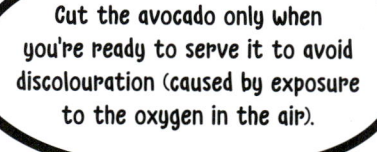

CLASSIC BEAN SALAD

MAKES 6 SERVINGS

1 14-oz can (398 ml) cut green beans, drained
1 14-oz can (398 ml) cut waxed beans, drained
1 19-oz can (540 ml) red kidney beans, washed and drained
1 medium onion, sliced thinly into rings
1 medium green pepper, cut into strips
½ cup white sugar
½ cup oil
1 cup white vinegar
Salt and pepper to taste

1. Mix the dressing ingredients – the sugar, oil, vinegar and salt and pepper.
2. Mix the beans, onion and green pepper in a large bowl
3. Add the dressing to the beans and vegetables and stir gently. Refrigerate for 24 hours before serving, stirring occasionally during that time.

Rule of thumb: This bean salad uses the acid from the vinegar to naturally wilt or sweat the onion and soften the peppers and other ingredients. The juices from these ingredients add further flavour to the beans and allow them to marinate (and preserve) in the resulting dressing liquid.

This is perfect BBQ or picnic salad as it self-preserves and lasts longer than a green salad..

GOURMET TANGY BEAN SALAD

MAKES 8–10 SERVINGS

1 cup (approximately 2 handfuls) green beans, trimmed
1 14-oz can (398 ml) cut waxed beans, drained
1 19-oz can (540 ml) red kidney beans, washed, drained
4 green onions, thinly sliced
2 large carrots, peeled, fine julienne sticks
½ cup frozen edamame, thawed and shelled
¾ cup red-wine vinegar
½ cup tomato juice
3 tbsp apple juice
¼ cup extra virgin olive oil
½ cup sugar
2 tsp Worcestershire sauce
2 tsp Dijon mustard
1 clove garlic, minced
2 dashes Tabasco

1. Cook green beans in large saucepan for 10 minutes, until they are tender. Strain and rinse the beans under ice cold water.
2. Mix all the dressing ingredients in an extra-large bowl. Add the canned beans, carrot, edamame and green onion. Refrigerate for 24–48 hours.
3. Add the green beans and toss in the salad just before serving so the colour of the green beans is preserved.

CLASSIC PICNIC POTATO SALAD

MAKES 6 SERVINGS

5 cups cooked Yukon Gold or new white potatoes, cubed
1 small onion, grated
2 tbsp fresh parsley, chopped
⅓ cup mayonnaise
¼ cup white vinegar
¼ tsp salt
Pepper to taste

1. While the potatoes are still warm, add the vinegar, salt, and grated onion.
2. Just before serving, add the mayonnaise and parsley. Adjust the sweetness if desired with a tablespoon of white sugar.

Rule of thumb: Potato salad with egg-based mayo may pose a health risk if it's brought to a picnic or BBQ where it can't be refrigerated.. Add more vinegar to protect against foodborne illness.

In adding more vinegar and lemon juice, the level of saltiness you perceive may become too intense and undesirable. Offset that with a little sugar to balance the flavours.

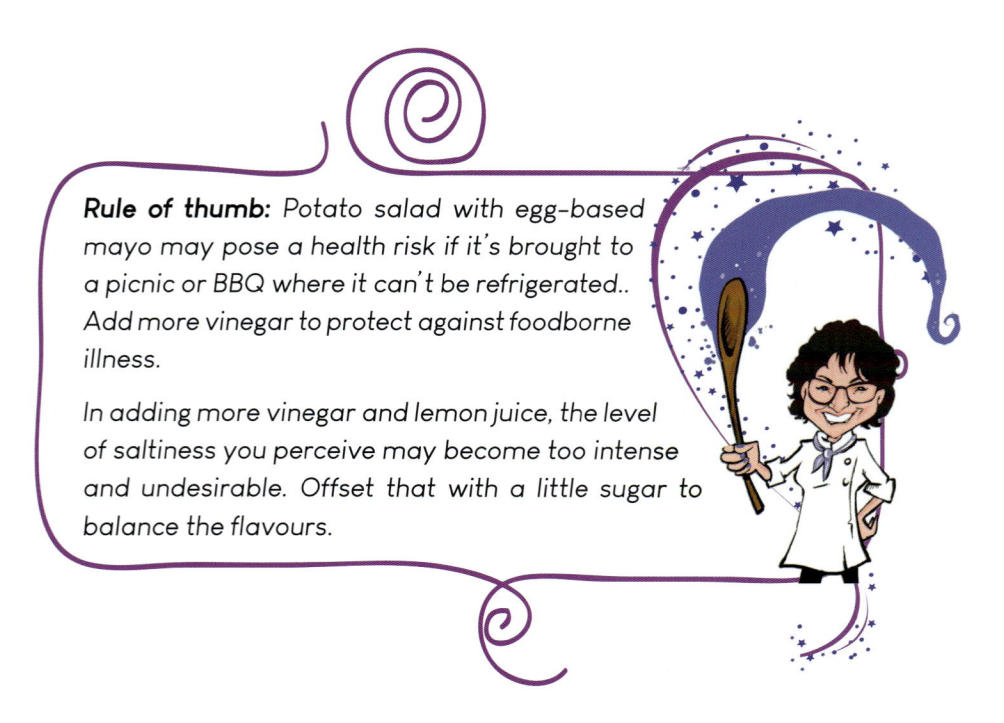

WARM GERMAN POTATO SALAD

MAKES 6 SERVINGS

5 cups potatoes, sliced ¼" thick and boiled
4–5 slices bacon, precooked, chopped
1 medium onion, grated
¼ cup fresh parsley, chopped
¼ cup beef broth
½ cup cider vinegar
1 tsp sugar
¼ tsp salt
1 cup mayonnaise
Pepper to taste

1. While the potatoes are still warm, add the vinegar, broth, salt, and grated onion. Stir to mix and coat the potatoes.
2. Add the chopped cooked bacon and stir it in.
3. Just before serving, add the mayonnaise and parsley. Adjust the sweetness if desired with a tablespoon of white sugar.

62

5. Stocks, Soups and Sauces

MY STORY: MAKING OF A SEAFOOD CHOWDER

When I first graduated from university, I worked for a small fish processor in a hamlet of St. John's. My first task was to develop a special seafood chowder that was derived from underutilized fish. At the time, skate, typically *Raga radiata,* was being incidentally caught in cod-fishing trawls without a use for them except for fish meal. In the skate wings is a wonderful piece of meat, a medallion of sweet flesh, likened to scallops. As well, the belly flap of the cod was also in abundance yet did not have a commercial use. The chowder I developed incorporated both byproducts along with a creamy flavourful base that included the stock from the fish trimmings and skate bones. I created creaminess using typical French cuisine sauce techniques, particularly a béchamel sauce using a roux of flour as the thickening agent as well as butter and milk but further charged with the protein-rich, gelatinous fish stock. This chowder is the one I have served to my family over the years and now to the thousands of people who come to our restaurant, Java Jack's Restaurant & Gallery, in Gros Morne, Newfoundland, Canada. The chowder is rich in nutrition and tradition and super big on flavour.

THE MAGIC OF SCIENCE: THICKENING

The magic of thickening soups and sauces and being able to enhance the glossy richness of soup and sauce is in the collagen from the stock and the release of complex carbohydrates of flour, cornstarch, and root vegetables.

Stock magic: Most bones and trimmings of carcasses contain collagen along with other distinct flavour components. Collagen is a tight, very strong network of protein that wraps around muscle bundles and is the main component of tendons, ligaments, and joints. When meat, trimmings, or bones are cooked or heated at high temperatures for extended periods of time, the collagen melts and pools at the bottom. The meat is tender and falls from the bones, and the bones fall away from each other. The residual liquid, consisting of melted collagen and moisture, appears translucent, sometimes opalescent. It is good to skim and strain out all the little bits for a richer broth. When it cools, the collagen firms the stock into a gel. The more concentrated the collagen, the firmer the gel. This is the base for most soup or sauce. It also creates a glossy sheen in the light and evenly coats the back of a spoon. It is often thought that these less-desirable cuts and bones are worthless, but they are the secret to amazingly good soups and sauces.

Starch magic: The major energy reserve of most plants is starch, most abundant in seeds, roots, and tubers. Starch is the only carbohydrate universally found in small packets called granules. These granules differ depending on the source; some are large and some small. Potato is the largest, and rice is the smallest. Each starch granule contains different levels of effective starch components i.e. amylose, and amylopectin. They are bound tightly together and housed by an outer shell. These starch granules do not typically dissolve or even suspend in cold water. In fact, the water and starch separate, and the starch settles out. However, when the temperature increases, the starch molecules start to vibrate more vigorously, allowing moisture to penetrate and unravel the starch core. With continued heating and an abundance of water, the starch granules swell and create viscosity. This is the gelatinization point and happens within a narrow temperature range. Starch gelatinization, the viscosity of starch solutions and the characteristics of the gels, depend also on the other constituents present. For soups and sauces, starch contained in flour, readily available cornstarch, starches in potato, rice, carrots and barley all provide an opportunity to create viscosity when the soup or sauce gets thick. The objective of starch is to hold moisture, create a gel matrix, and add body (thickening and viscosity) to any soup, sauce, or gravy.

STOCKS

Stock and broth are close kin of each other. Broth is created with meat, bones, and vegetables and typically does not gel when cold. The objective of stock, on the other hand, is to be clear and fat free and solidify into gel when cold. Sometimes, bones can be roasted to impart additional flavour. The ingredients for a stock need not be prime, but the stockpot should not be used as a catchall for leftovers.

- Chicken Stock
- Fish Stock
- Beef Stock

Rule of thumb: The hallmark of a good stock or broth is clarity achieved by simmering rather than boiling the liquid. Let the pot come to a boil for only a minute or two and then turn it down to a simmer.

Skim off any foam or scum that rises to the surface to help with clarity.

Strain the final stock through a fine sieve, removing all the little bits. Chill and remove the fat layer before use.

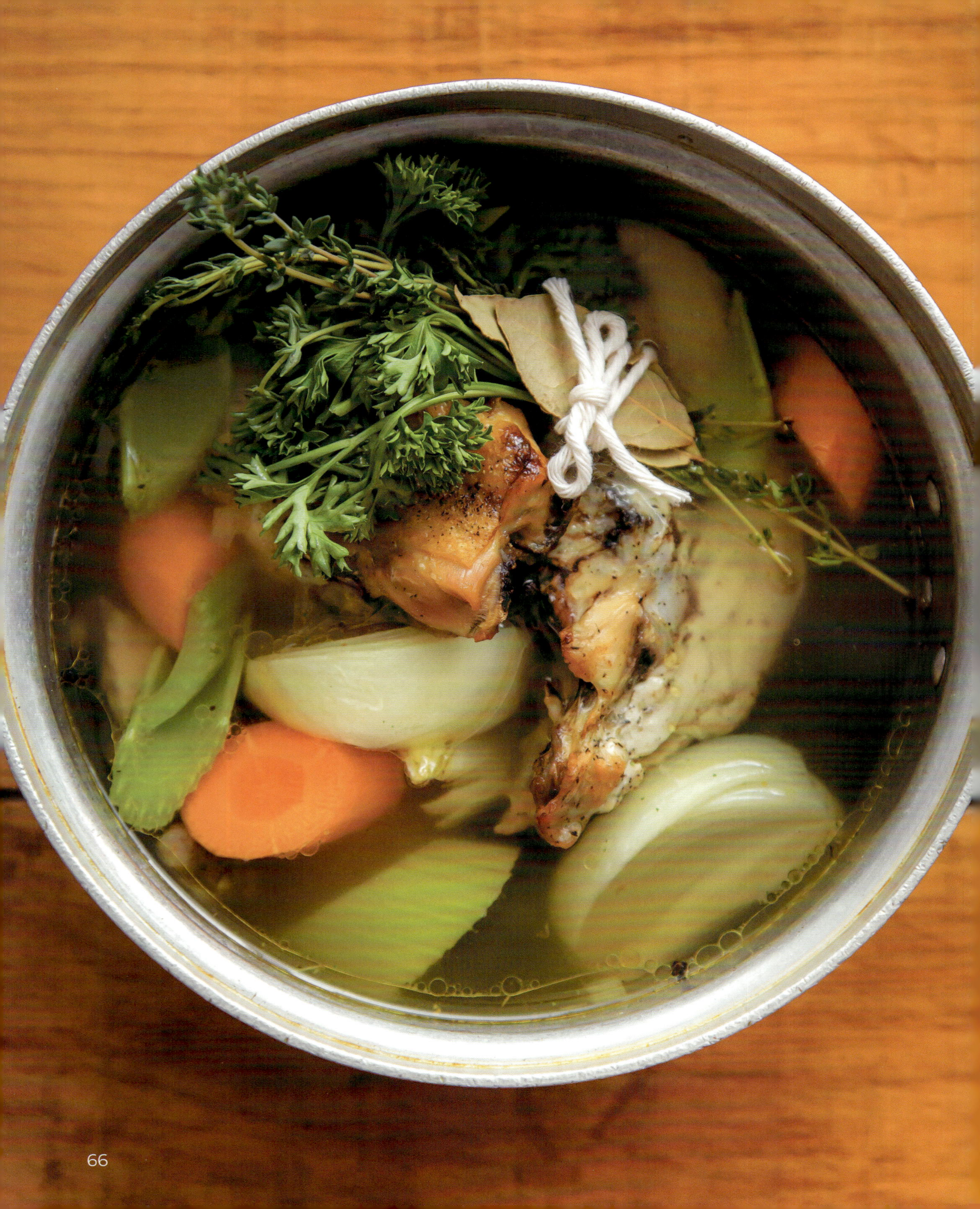

CHICKEN STOCK

MAKES ABOUT 8-12 CUPS

Chicken bones (backs, necks, carcass from 3-lb (~1 ½ kg) chicken)
2 medium onions, peeled and quartered
2 medium carrots, peeled and quartered
2 stalks of celery, cut into 2" (5 cm) pieces
10 peppercorns
1 clove garlic
Bouquet garni
12–16 cups water

BOUQUET GARNI includes sprigs of thyme, parsley, and a bay leaf tied together with string or wrapped in cheesecloth.

1. Put the bones and vegetables in a stockpot. Add the bouquet garni, peppercorns, garlic, and water.
2. Bring slowly to a boil, skimming often. Simmer for 3–4 hours.
3. Remove the bouquet garni and discard. Strain and chill. Skim off any fat.

Rule of thumb: If you use a carcass and bones from a cooked chicken or turkey, remove any obvious fat. Follow the recipe above.

Opposing popular belief, skin is not all fat. It contains about ⅓ fat, but the other ⅔ is protein, mostly collagen, and moisture. It adds good nutrition. So add the skin to the stockpot!

FISH STOCK

MAKES ABOUT 4 CUPS

1 tbsp butter
1 medium onion, finely chopped
1 ½ lb fish bones and trimmings (no skin)
1 cup white wine or juice from ½ lemon
4 cups water
10 peppercorns
Large bouquet garni

> **LARGE BOUQUET GARNI** includes a few sprigs of thyme, a bunch of parsley, and a few bay leaves tied together with string or wrapped in cheesecloth.

1. In a large stockpot, melt the butter and cook the onions slowly until soft, not brown, 7–10 minutes.
2. Add the bones, wine, water, and peppercorns. Add the bouquet garni.
3. Bring slowly to a boil, skimming often. Simmer uncovered for 20 minutes to 1 hour.
4. Remove the bouquet garni and discard. Strain and chill.

Rule of thumb: No skin! While skin has a lot of collagenic material and great gelling properties, it imparts a dark colour to the stock.

Onions add good flavouring for fish stock. Carrots are too sweet and can tint the stock.

BEEF STOCK

MAKES ABOUT 8-12 CUPS

Beef bones, cut into pieces (about 2 kg)
2 medium onions, peeled and quartered
2 medium carrots, peeled and quartered
2 stalks of celery, cut into 2" (5 cm) pieces
10 peppercorns
1 clove garlic
Bouquet garni
1 medium tomato, chopped, or 1 tbsp tomato paste
12–16 cups water

> You can use bones from large animal game, such as moose or caribou, as well as lamb and veal bones.

1. Put the bones and vegetables in a stockpot. Add the bouquet garni, peppercorns, garlic, tomato, and water. For more flavour, brush the bones with tomato paste and roast in a hot oven first before putting them in the stockpot with the vegetables.
2. Bring slowly to a boil, skimming often. Simmer for 3–4 hours until about ¾ original volume.
3. Remove the bouquet garni and discard. Strain and chill. Skim off any fat.

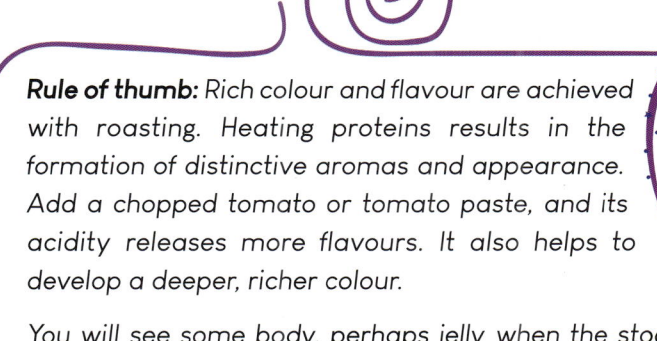

Rule of thumb: Rich colour and flavour are achieved with roasting. Heating proteins results in the formation of distinctive aromas and appearance. Add a chopped tomato or tomato paste, and its acidity releases more flavours. It also helps to develop a deeper, richer colour.

You will see some body, perhaps jelly when the stock cools. It is the gelatin formed from melting the collagen. Good nutrition!

BROTH SOUPS

You can further enhance the richness of stocks by adding other flavourful ingredients. For example, adding onions, beer, and herbs to a traditional French onion soup, or orange, seafood, and fennel to a bouillabaisse, or chicken, pasta, and diced vegetables to base stock will elevate the flavour, nutrition, and experience.

- Chicken Noodle Soup
- French Onion Soup with Cheddar and Ale
- Bouillabaisse

Rule of thumb: Once the stock is made, heat and simmer, rather than boil, your soup and let the vegetables and other ingredients cook. Vigorously boiling can develop unwanted off flavours.

CLASSIC CHICKEN NOODLE SOUP

MAKES 6 SERVINGS

1 medium onion, peeled and diced in ½" (1cm) chunks
1 tbsp butter
½ cup carrots, diced or uniformly sliced
½ cup celery, sliced
½ cup frozen green peas

1 cup roasted chicken, cubed
1 tsp Italian seasoning
4 cups chicken stock
1 tbsp fresh parsley, finely chopped
100 g spaghetti noodles, cracked in half
Salt and pepper to taste

1. In a large saucepan, melt butter over medium heat. Add the onions and carrots and cook until they are soft and translucent, about 5 minutes.
2. Add chicken and green peas and heat them through. Stir in the Italian seasoning.
3. Stir in stock. Bring to a boil, then reduce heat. Cover and simmer for 5–10 minutes to develop flavours.
4. Add the dry pasta noodles and stir. Let simmer for another 10 minutes until the noodles are soft.
5. Add parsley and stir. Taste the broth and season with salt and pepper before serving.

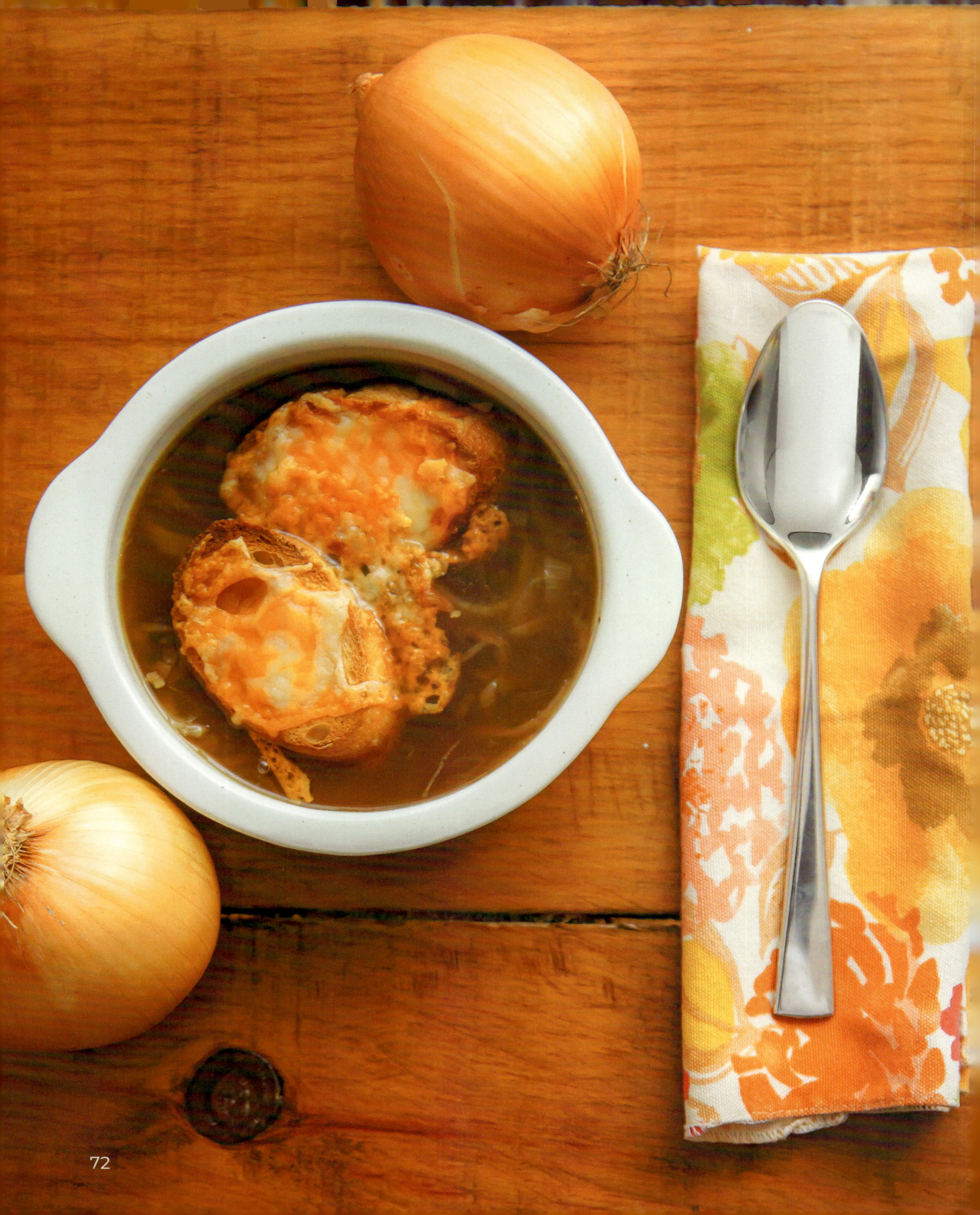

FRENCH ONION SOUP WITH CHEDDAR AND ALE

MAKES 6 SERVINGS

4 large onions, peeled and sliced
2 tbsp butter
1 cup beer, preferably ale
¾ tsp dried thyme leaves
½ tsp sugar
½ tsp Worcestershire sauce

4 cups beef stock
12 slices baguette bread
1 ½ cups old cheddar or Gruyère cheese
⅓ cup grated parmesan cheese
⅓ cup sour cream
Salt and pepper to taste

1. Slice the onions about ¼ inch thick. In a large, wide saucepan, melt butter over medium heat. Add the onion slices and cook them, separating into rings and stirring frequently, until browned, about 10 minutes.
2. Add beer, thyme, sugar, and Worcestershire sauce. Increase the heat to medium high and bring to a boil uncovered. Stir the mixture often until the liquid is reduced by half, about 5 minutes.
3. Stir in broth. Bring everything to a boil, reduce the heat, then cover and simmer for 5–10 minutes to develop the flavours.
4. Season with salt and pepper to taste. (You can prepare the recipe to this point and refrigerate it for up to five days.) Skim any foam that appears on the surface of the pot often.
5. Cut a baguette into ½" slices. Place the slices on a baking sheet and toast them lightly in the oven.
6. Meanwhile, in a bowl, stir together the cheddar and parmesan cheese and sour cream.
7. Ladle the soup into oven-proof bowls, leaving space for toast. Place on a baking sheet.
8. Spread the cheese mixture over the toast, and float two pieces on each bowl of soup.
9. Bake in 450°F oven until the cheese is golden and soup bubbles, about 13 minutes. Serve immediately.

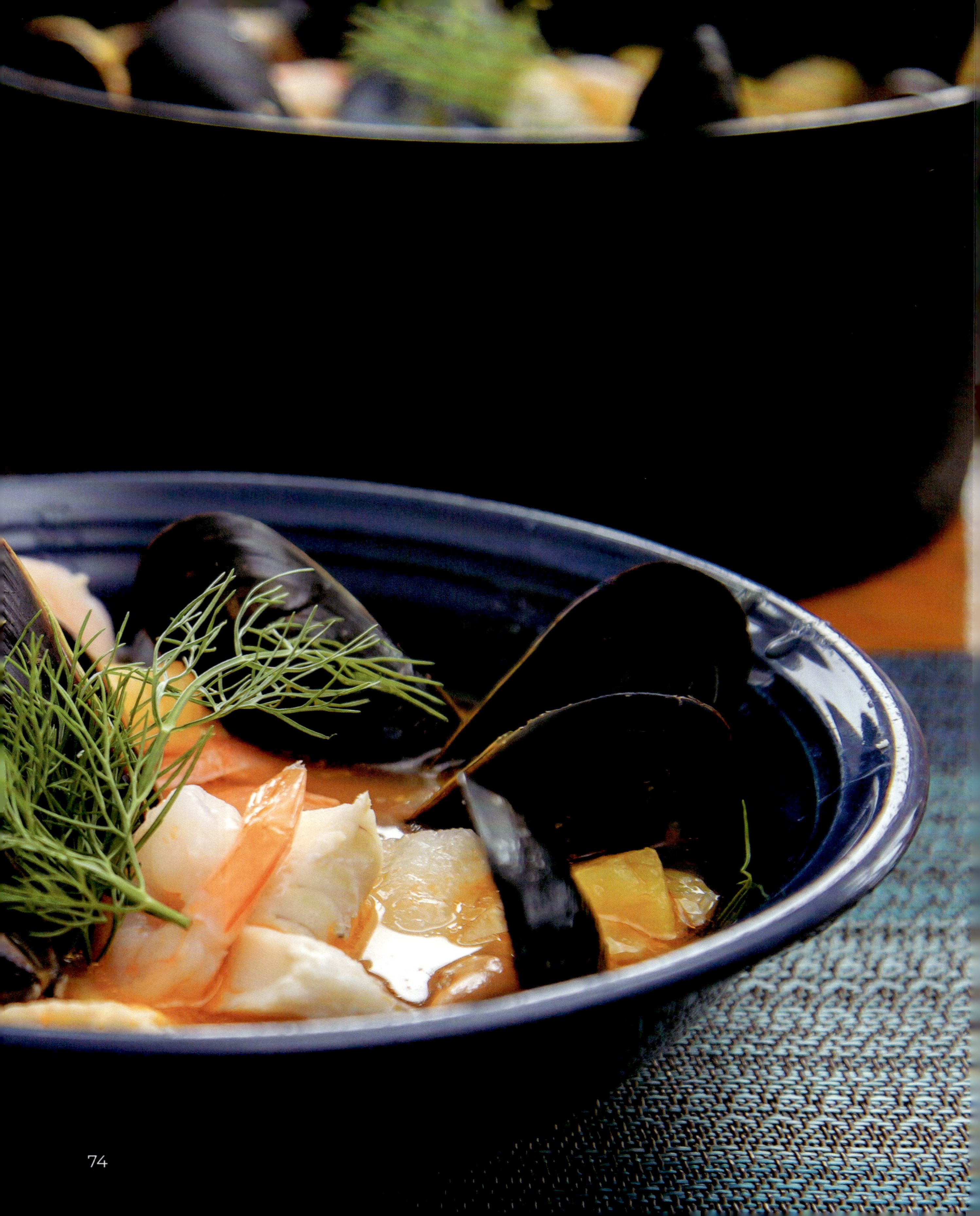

BOUILLABAISSE

MAKES 6–8 SERVINGS

½ cup olive oil
1 bulb fennel, stems trimmed, cored, and sliced
1 large onion, peeled and sliced
1 small leek, white and light green parts only, sliced
1 large tomato, peeled, seeded, and chopped
3 garlic cloves, peeled and chopped
1 tbsp tomato paste
Cayenne pepper to taste
1 tsp saffron threads
2 medium russet potatoes, peeled, diced
6 cups fish stock
1 bouquet garni
Salt and fresh ground pepper to taste
1 lb (450 g) peeled shrimp
1 lb (450 g) cod, halibut, or other flaky white fish, cut into large chunks
1 lb (450 g) mussels, scrubbed and debearded
2 strips of orange zest
Juice from a large orange

1. In a large, heavy-bottomed stockpot, heat the olive oil over medium-high heat.
2. Add the fennel, onions, and leeks, and cover with a tight-fitting lid. Cook the vegetables, stirring occasionally, until just they just begin to tenderize, about 5 minutes.
3. Add the tomato and garlic. Cover, and continue cooking until the tomatoes begin to break down, 2–3 minutes.
4. Add the tomato paste, cayenne, saffron, potatoes, fish stock, and bouquet garni. Add the orange juice and orange zest. Season with salt and pepper and bring the liquid to a boil.
5. Cover the pot, reduce the heat to medium, and simmer until the potatoes are cooked, 10–15 minutes.
6. Add the seafood. Simmer the soup, stirring very minimally to preserve the whole pieces of seafood, until the shellfish have opened and the fish fillets have cooked through, 5–10 minutes depending on the seafood being used.
7. To serve, remove and discard the bouquet garni and the orange zest. Carefully remove the fish and shellfish, arranging them in warm bowls. Ladle the broth and potatoes over the fish and garnish with herbs such as dill, fennel fronds or parsley. Serve warm.

Rule of thumb: Add the seafood in order of firmness and cooking time, with the denser fillets of fish going into the pot first, followed by shellfish, and last shrimp and scallops.

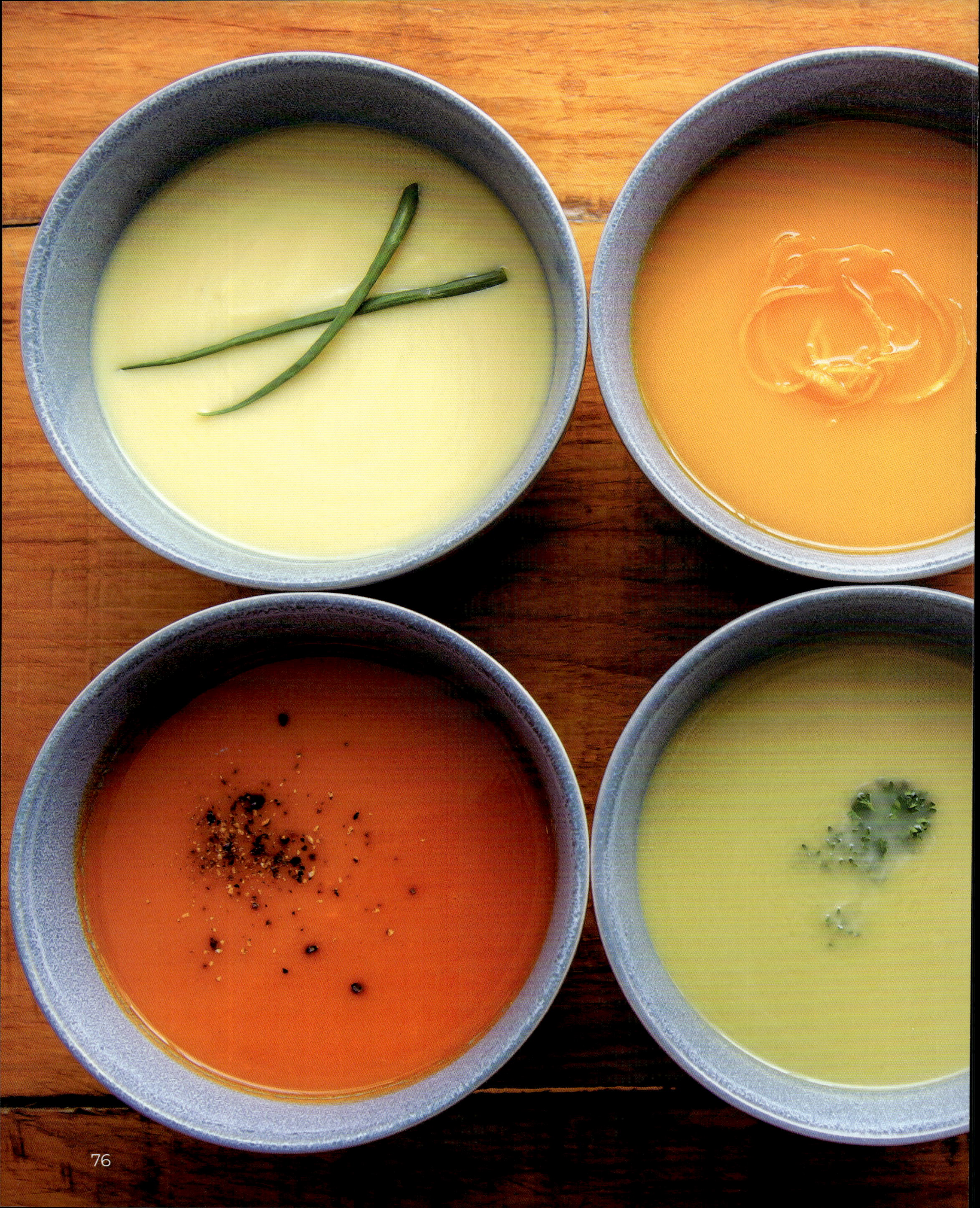

VELVETY SMOOTH SOUPS

I love soups in summer or winter and am particularly partial to smooth, creamy soups. The base for the flavour comes from the stock, just as a canvas is to an artist or the primer is to a painter. I love to see how the magic of grains or starchy root vegetables can create viscosity as the cooking process advances. Further viscosity develops when the cellular starch is released as the blender pulverizes the mixture into a velvety-smooth soup.

- Classic Cream of Any Vegetable Soup
- Vichyssoise (Classic Potato and Leek Soup)
- Pumpkin Ginger Orange Soup
- Roasted Red Pepper Soup

Heat releases and activates the starch in potato and carrots. It also activates the pectin in the flesh and skin of apples for additional support in developing a thick, velvety-smooth texture. Potato is my one ingredient of choice as it is a bland-tasting vegetable and does not distract from the taste of the real star of the dish. Apple imparts a background sweetness, but unless apples are used in large quantities, the flavour contribution is minimal. Carrots have a distinct colour and flavour and complement many other "orangey" ingredients. Oranges complement carrots and offer pectin from the pulp.

Rule of thumb: *Cook the potatoes, carrots, and starchy vegetables until very soft and mushy so the resulting soup has body. Strain the final soup through a fine sieve to remove any pulp or fibres.*

CLASSIC CREAM OF ANY VEGETABLE SOUP

MAKES 6 SERVINGS

1 medium onion, peeled and finely chopped
1 medium potato, peeled and diced (2 cm)
1 medium apple, unpeeled and diced
2 tbsp butter
4 cups chicken or vegetable stock
2–3 cups vegetables, cleaned and cut into bite-size pieces, including leaves, stalks, and florets
Salt and white pepper

Vegetables: celery, broccoli, leeks, carrots, sweet potatoes, pumpkin, green peas, mushrooms

1. In a large saucepan, melt the butter over medium heat. Add the onions and cook until they are soft and translucent, about 2 minutes.
2. Add the potatoes, diced apple, and any edible stalks. Stir and cook for another minute to release the starch.
3. Add the vegetables and heat them through to brighten the colour.
4. Stir in the stock. Bring to boil, reduce heat, then cover and simmer for 10–15 minutes (maximum 30 minutes).
5. Using a blender, puree the soup until very fine, smooth, and velvety.
6. Strain the pureed soup through a medium mesh strainer to catch any unblended big pieces.
7. Taste the soup and season with salt and pepper.

Optional: after blending and straining the soup, add ¼–½ cup heavy cream or plain yogurt, and heat the soup thoroughly before serving.

Rule of thumb: You can add the edible stalks of vegetables with the onions and potatoes. The high-humidity heat of the cooking process breaks down the tough fibres (cellulose), and the blending process pulverizes them to help create body for the soup.

Additionally, if you cook the leaves and florets of the vegetables before you add the stock, the colour brightens and sets, creating a fresher-looking soup.

PUMPKIN GINGER ORANGE SOUP

MAKES 6 SERVINGS

1 medium onion, peeled and finely chopped
2 medium carrots, peeled and diced (2 cm)
2 tbsp butter
3 cups chicken or vegetable stock
1 ½ cups cooked or canned pumpkin
1 tbsp grated ginger
1 tsp curry powder
Zest, grated from half an orange
Juice from one orange
1 cup full-fat coconut milk
Salt and white pepper to taste
Green onion (optional)

1. In a large saucepan, melt the butter over medium heat. Add the onions and cook them until they are soft and translucent, about 2 minutes.
2. Add the carrots, stir, and cook for another minute to release the starch.
3. Add the pumpkin, ginger, orange juice, and zest.
4. Stir in the stock. Bring to boil, reduce heat, then cover and simmer for 10–15 minutes (maximum 30 minutes).
5. Using a blender, puree the soup until smooth and velvety.
6. Strain the soup and return the soup to the saucepan, add the coconut milk, and incorporate it into the soup with a whisk. Cook over low heat until the soup is just hot. Do not boil.
7. Taste the soup and season with salt and pepper. Garnish with green onion and freshly cracked pepper.

ROASTED RED PEPPER SOUP

MAKES 6 SERVINGS

3 roasted red peppers or one 7-oz jar roasted red bell peppers, drained
1 tbsp olive or canola oil
1 small onion, chopped
1 clove garlic, chopped
2 ½ cups chicken or vegetable stock
1 ½ tsp sugar
2 tbsp fresh basil leaves, chopped, or 2 tsp (10 mL) dried basil plus more fresh basil for garnish
¾ cup plain yogurt (optional)
Salt and pepper to taste

1. In a three-quart saucepan, heat the oil over medium heat until hot. Add the onion and garlic and cook, stirring occasionally for about 5 minutes until the vegetables are tender.
2. Add the broth, roasted peppers, sugar, salt, and pepper. Cook uncovered 10 minutes over medium heat, stirring occasionally.
3. Place half the mixture with half the basil in a blender. Cover and puree until smooth. Repeat. Return to the saucepan.
4. Strain the soup. Stir in the yogurt, if using and heat over low heat until just hot. Do not boil. Garnish with additional fresh basil and freshly ground black pepper if desired.

HOW TO ROAST RED PEPPERS
Preheat the oven to 400°F. Place the red peppers on a baking sheet lined with parchment paper. Roast in the oven until the skin starts to blister. There will be black char spots on the skin, and the pepper will look soft and soggy. Remove the peppers from the oven. Using tongs, put the roasted red peppers in a large bowl and seal with plastic wrap. doing this creates humidity so the peppers' skin lifts easily once cooled. Remove the peppers' skin and cut in half. Remove all the ribs, stems, and seeds and discard.
Voilà! Roasted red peppers.

CLASSIC POTATO AND LEEK SOUP (VICHYSSOISE)

MAKES 6 SERVINGS

1 medium onion, peeled and finely chopped
3 leeks, cleaned and finely sliced up to the dark green portion
2 tbsp butter
3 potatoes, peeled and diced (2 cm)
4 cups chicken stock
¼ tsp mace
Salt and white pepper
For added richness, add 1 cup heavy cream (35% fat)

1. In a large saucepan, melt the butter over medium heat. Add the onions and leeks and cook them until they are soft and translucent, about 2 minutes.
2. Add the potatoes, stir, and cook for another minute.
3. Stir in the stock. Bring to boil, reduce the heat, then cover and simmer for 10–15 minutes.
4. Using a blender, puree the soup until very fine, smooth, and velvety.
5. Strain the soup through a medium mesh strainer and return the soup to the saucepan.
6. Taste the soup and season with salt and pepper.
7. Optional: Add heavy cream and heat the soup thoroughly before serving.

SAUCES

In French cuisine, five mother sauces are the culinary base for any kind of soup or sauce you might want to make. The most used of these mother sauces in my repertoire are béchamel, velouté, and hollandaise. Hollandaise sauce uses egg yolk for viscosity. An egg has its own amazing properties (see Chapter 6). Both béchamel and velouté use flour, which contains starch for viscosity development and wheat gluten as an emulsifier with fat. The combination is magical in that wheat flour is used the world over for cooking. While many substitute with one or more ingredients, honestly there is no true comparison for wheat flour's magic.

- Sauce Base: Making a Roux
- Béchamel Sauce Two Ways
- Cheese Sauce (Mornay), Mushroom Cream Sauce, Curry Sauce
- Mac and Cheese
- Cod au Gratin
- Velouté and Gravy
- Murphy's Seafood Chowder
- Espagnole Sauce
- Authentic Shrimp Creole

The Magic in Making a Sauce

Simply create a roux with butter (source of fat) and flour (source of protein and starch).

Heat to melt the fat in the butter and release the protein and starch in the flour.

Add some liquid, such as stock, milk, or water and stir. Allow the thickening to occur.

Add some seasonings to taste.

A thin sauce with more liquid added proportionally is useful to accompany meat or creating a chowder. It sticks to the back of the spoon.

A thicker sauce coats pasta and vegetables.

SAUCE BASE: MAKING A ROUX

½ cup all-purpose flour
½ cup butter or fat

The starting point for common sauces is a flour-and-butter base called a roux. This is French for *russet brown*. Classic French teachings state a one-to-one ratio by volume of flour and butter. Melt the butter before adding the same amount of flour.

Note that different flours have different protein and starch percentages. Flour that contains more protein, such as bread flour, holds more fat. Flour that contains more starch, like cake flour, holds more water. So stick to all-purpose flour with 11% protein.

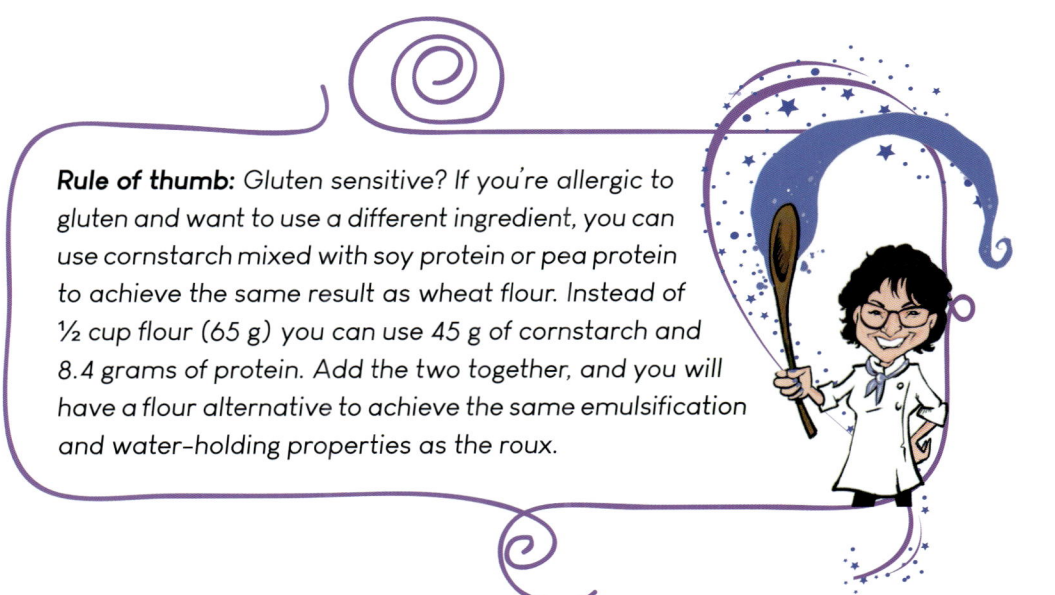

Rule of thumb: Gluten sensitive? If you're allergic to gluten and want to use a different ingredient, you can use cornstarch mixed with soy protein or pea protein to achieve the same result as wheat flour. Instead of ½ cup flour (65 g) you can use 45 g of cornstarch and 8.4 grams of protein. Add the two together, and you will have a flour alternative to achieve the same emulsification and water-holding properties as the roux.

BÉCHAMEL SAUCE TWO WAYS

MAKES 1 CUP

FIRST WAY

½ medium onion, peeled and diced (1 cm)
1 ½ tbsp butter
1 ½ tbsp flour
1 cup milk
Pinch nutmeg
Salt and white pepper

1. Melt the butter in a medium saucepan. Add the onions and cook until the onions are soft and translucent.
2. Add the flour and cook until foaming, about 1 minute.
3. Whisk in the milk quickly. Heat the mixture until it is thickened.
4. Pass through a strainer. Return the sauce to the heat and season with salt and white pepper to taste.

SECOND WAY

1 ½ tbsp butter
1 ½ tbsp flour
1 cup milk
A few slices onion
1 bay leaf
½ tsp peppercorns
Pinch nutmeg
Salt and white pepper

1. Add the onion, bay leaf, and peppercorns to a small saucepan that contains the milk. Bring to a boil, remove from the heat, cover, and set aside to let the flavours infuse the milk.
2. Melt the butter in a medium saucepan. Add the flour and cook until foaming, about 1 minute, creating a roux.
3. Strain the milk into the roux and whisk it in quickly. Heat the mixture until it is thickened.
4. Pass again through a strainer to eliminate any lumps. Return the sauce to the heat and season with salt and white pepper to taste.

Rule of thumb: To thicken the béchamel sauce for gratins, reduce the amount of milk or increase the roux.

To make the béchamel thinner for chowders or soup, add more liquid such as water, milk, or stock.

Keep an eye on the milk. It curdles and separates when cooked too hot. It also burns to the bottom if not stirred on occasion.

85

FREESTYLE
CHEESE SAUCE (MORNAY)

MAKES 1 ½ CUPS

1 cup béchamel sauce
½ cup Gruyère cheese, grated
½ cup parmesan cheese, grated
1 tsp Dijon mustard

> Different cheeses such as old cheddar, Gruyère or Jarlsberg add nice flavours. Avoid mozzarella as it will make the sauce stringy and is limited in flavour delivery.

1. Warm up the béchamel sauce to almost boiling.
2. Take the sauce off the heat and add the cheeses and Dijon mustard. Stir until melted. Season with salt and pepper.

MUSHROOM CREAM SAUCE

MAKES 1 ½ CUPS

1 cup béchamel sauce
½ cup heavy cream (35% whipping cream)
1 package (250g) white or brown cremini mushrooms, sliced thin

1. Sauté the mushrooms in butter in a saucepan.
2. Dab the mushrooms with a paper towel to absorb the excess fat.
3. Add the béchamel sauce. Bring to a boil, add the heavy cream and reduce the heat. Stir, and season with salt and pepper.

CURRY SAUCE

MAKES 1 ½ CUPS

2 cups béchamel sauce
1 medium onion, finely chopped
2 tbsp butter
1 tbsp curry powder
1 medium tomato, peeled, seeded, and chopped

1. Sweat onions in butter.
2. Add the curry powder and chopped tomato. Stir and cook for another 2–3 minutes.
3. Add the béchamel sauce. Simmer until reduced to 1 ½ cups.
4. Strain and season with salt and pepper.

MAC AND CHEESE

MAKES 4 SERVINGS

1 cup dry macaroni elbows
4 cups boiling water
1 tsp salt
1 cup béchamel sauce
1 tsp Dijon mustard
½ cup old cheddar cheese
¼ cup parmesan cheese
¼ cup breadcrumbs
Salt and pepper to taste

1. Preheat the oven to 400°F.
2. Add macaroni elbows to the salted boiling water. Cook for 10 minutes until al dente. Strain the water.
3. Mix together the béchamel sauce, Dijon mustard, ½ cup old cheddar, and parmesan cheese
4. Add cooked macaroni and coat with cheese sauce. Season with salt and pepper.
5. Spoon out the mac and cheese into a 9x11 baking dish. Top with a mixture of ½ cup cheddar and breadcrumbs.
6. Cook at 400°F for 15–20 minutes until cheese is melted and sauce is bubbling.

COD AU GRATIN

MAKES 4 SERVINGS

Every Newfoundlander has a recipe for this dish. It was made popular when it was showcased in the *Come From Away* Broadway production. I have added some distinctive cheese such as Gruyère and a little Dijon mustard to enhance the cheesy flavour. Certainly, using just the regular cheddar cheese in your fridge will still make a wonderful dish your family will love.

2 ½ cups béchamel sauce (see p. 84)
½ cup Gruyère cheese, grated
1 tsp Dijon mustard
1 ½ cup cheddar cheese, grated
1 lb cod fillets, cut into chunks
½ cup fine breadcrumbs
2 tbsp fresh parsley (2 tsp dried parsley)
Salt and freshly cracked black pepper to taste

1. Preheat the oven to 400°F.
2. Place the béchamel sauce in a large saucepan and heat.
3. Add the Gruyère cheese and ½ cup cheddar cheese. Stir and melt the cheese.
4. Add the cod and stir gently until it is just cooked.
5. Taste and season with salt and ample cracked black pepper. Portion into a baking dish or individual ramekins.
6. In a bowl, mix together the breadcrumbs, remaining cheddar, and parsley.
7. Add this crumb mixture generously to the top of each ramekin or the entire baking dish. Bake for 10–15 minutes until the cheese is melted and the sauce is bubbling.

Rule of thumb: Mornay cheese sauce is the basis for this dish. You can make it to pour over steamed veggies or potatoes. It consists of béchamel sauce, Gruyère cheese, Dijon mustard and aged cheddar cheese. Add other cheeses you like as well.

VELOUTÉ

MAKES 1 ½ CUPS

Velouté is white cream sauce made from a roux and stock but thinner at the start and cooked to develop the flavour.

1 ½ tbsp butter
1 ½ tbsp flour
1 ½ cup chicken, beef, fish, or vegetable stock
Salt and pepper to taste

1. Melt the butter in a medium saucepan.
2. Add the flour and cook until foaming for about 1 minute (or longer if a golden to brown colour is desired).
3. While whisking, add the stock until it is slightly thickened and covers the back of a wooden spoon.
4. Pass the sauce through a fine mesh strainer so you have silky smooth sauce.
5. Return the velouté to the heat and season with salt and pepper to taste.

GRAVY

Gravy is made with meat juices or drippings in the bottom of the roasting pan. The fat from the drippings is mixed with flour. Stock is added once the brown colour is developed (no need for a gravy browning). You might think about adding coffee for colouring as they do in Europe for ham. Add stock, wine, or water to thin a thick gravy.

MURPHY'S SEAFOOD CHOWDER

MAKES 6 SERVINGS

This is the famous seafood chowder we serve at Java Jack's Restaurant & Gallery in Rocky Harbour, Newfoundland, in the service community to Gros Morne National Park. It is wildly popular with our guests from far and wide. It has even appeared in a few local publications. Enjoy!

1 cup carrots, diced
1 cup turnip, diced
1 cup celery, diced
1 lb cod fillets or any white fish, fresh or frozen and thawed, skinned, boneless, cut into 1 ½" cubes
½ lb lobster meat and scallops, cut approximately ½" x ½", or any other combination of any seafood (mussels, lobster, scallops, shrimp) cut approximately ½" x ½"
3 cups thick béchamel sauce (see p. 84)
2 tbsp dried tarragon
Salt and pepper to taste

1. Add 3 cups of water to a saucepan and add the vegetables. Bring to a boil, then reduce heat and cook over medium heat until al dente (still firm, not soft), approximately 10–15 minutes.
2. When cooked, strain the vegetables and set aside the vegetables and reserve the vegetable water.
3. Add the cubed fish to another saucepan and add water just to the level of the fish. *Do not cover the fish with water, as the fish will create its own liquid during cooking.*
4. Cook gently so the fish is just cooked but not falling apart, about 5 minutes.
5. When cooked, strain the fish and retain the fish liquid to thin out the béchamel sauce. Set both aside.
6. Add the béchamel sauce to a large saucepan and heat to boiling, stirring often.
7. Add the fish liquid. Typically, use all the fish liquid. Then add enough of the vegetable water until sauce is a creamy texture.
8. Add the vegetables and fish and the lobster and scallops. Stir gently.
9. Add the tarragon and some salt and pepper to taste. Simmer for 5–10 minutes before serving.

For more on this seafood chowder check out the "How To" video on YouTube.

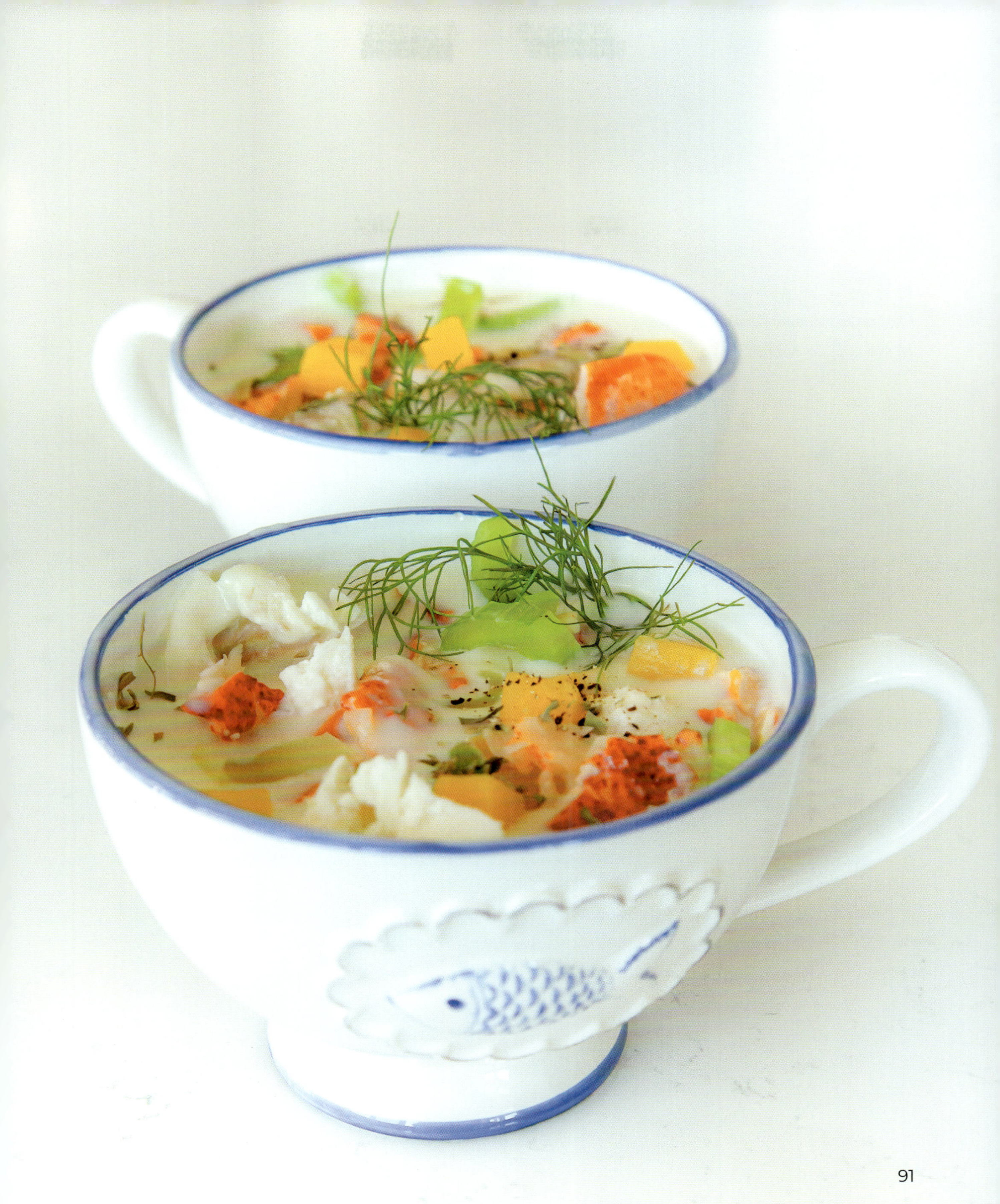

ESPAGNOLE SAUCE

MAKES 2 CUPS

The base for an espagnole sauce, which is considered to be one of the mother sauces of French cuisine, is a velouté (see p. 89). The roux is cooked thoroughly so it browns without scorching. As with most complex sauces, the sauce is a little harsh at first, but as it simmers it mellows and the flavour develops. The sauce darkens during cooking and takes on a brilliant glossy glaze.

3 tbsp oil
2 slices bacon, diced
½ medium onion, diced
½ medium carrot, diced
¼ cup flour

4 cups beef stock heated to be just warm
Bouquet garni (see p. 69)
1 medium tomato, quartered
2 tsp tomato paste
Salt and pepper to taste

1. Heat the oil in a pan. Add the bacon and cook until all the fat is rendered out of the bacon.
2. Add the carrot and onions and sauté until soft.
3. Add the flour and cook gently until the flour is dark brown, about 5 minutes.
4. Whisk in ¾ of the warmed stock. Whisk constantly until boiling and the sauce thickens.
5. Add the bouquet garni, tomato, and tomato paste.
6. Let simmer gently for 3–4 minutes, skimming often. Add the rest of the stock during the cooking.
7. Simmer the sauce until it's reduced by half. The sauce is done when it is glossy and rich.
8. Pass the sauce through a fine strainer and season with salt and pepper.

HOW TO MAKE A DEMI-GLACE
Mix 1 cup sliced sautéed mushrooms, 1 tsp tomato paste, ½ cup stock, 1 cup of espagnole sauce, and 1 tbsp sherry or port wine. Simmer and reduce the sauce to 1 cup. Remove from the heat and whisk in 1 tbsp butter for additional gloss.

AUTHENTIC SHRIMP CREOLE

MAKES 4 SERVINGS

I first had this dish in New Orleans, Louisiana, when my husband and I attended a cooking school demonstration. This recipe uses the fundamentals of making a sauce and, with seasonings and other inclusions, creates an OMG! masterpiece. Freestyle by substituting shrimp with chunks of cod, poaching them in the sauce to create Cod Creole.

1 cup onions, diced ¼-inch pieces
½ cup celery, diced
½ cup green peppers, cored and diced
2 tsp garlic, minced
1 tbsp butter
1 cup diced tomatoes
2 tbsp tomato paste
1 ½ cup chicken or vegetable stock
1 ½ tbsp brown sugar
2 bay leaves
3 tbsp creole seasoning
½ lemon, thinly sliced
2 tbsp butter
2 tbsp flour
1 ½ lbs uncooked shrimp, tails on
¼ cup parsley, chopped fine
¼ cup green onions, thinly sliced
Tabasco and salt to taste

CREOLE SEASONING

MAKES 2 CUPS

2 tbsp onion powder
2 tbsp garlic powder
2 tbsp dried oregano
2 tbsp dried basil
1 tbsp dried thyme
1 tbsp ground black pepper
1 tbsp ground cayenne pepper
5 tbsp paprika
3 tbsp salt

Mix all together and store is a sealed jar for up to six months.

1. Melt the butter in a frying pan or pot. Add the diced onions, celery, green pepper, and garlic and cook until the celery is soft and the onions are translucent.
2. Add the tomatoes, tomato paste, stock, bay leaves, brown sugar, Creole seasoning, and lemon slices. Heat to a boil and then turn down the heat to a simmer and cook for 20–30 minutes. Remove the bay leaves and lemon slices.
3. Melt the butter in a saucepan. Add the flour and cook slowly. Stir occasionally until the roux develops into a caramel-brown colour. Stir the roux into the other mix.

4. Let the mix thicken, then add the shrimp.
5. Adjust the sauce with some chicken or vegetable stock if it is too thick.
6. Add the shrimp. Cook on medium low until the shrimp are cooked. For Cod Creole, substitute the shrimp with chunks of cod fish.
7. Add the green onions and parsley. Simmer for 5 minutes. Taste and adjust with salt and Tabasco sauce. Serve with rice.

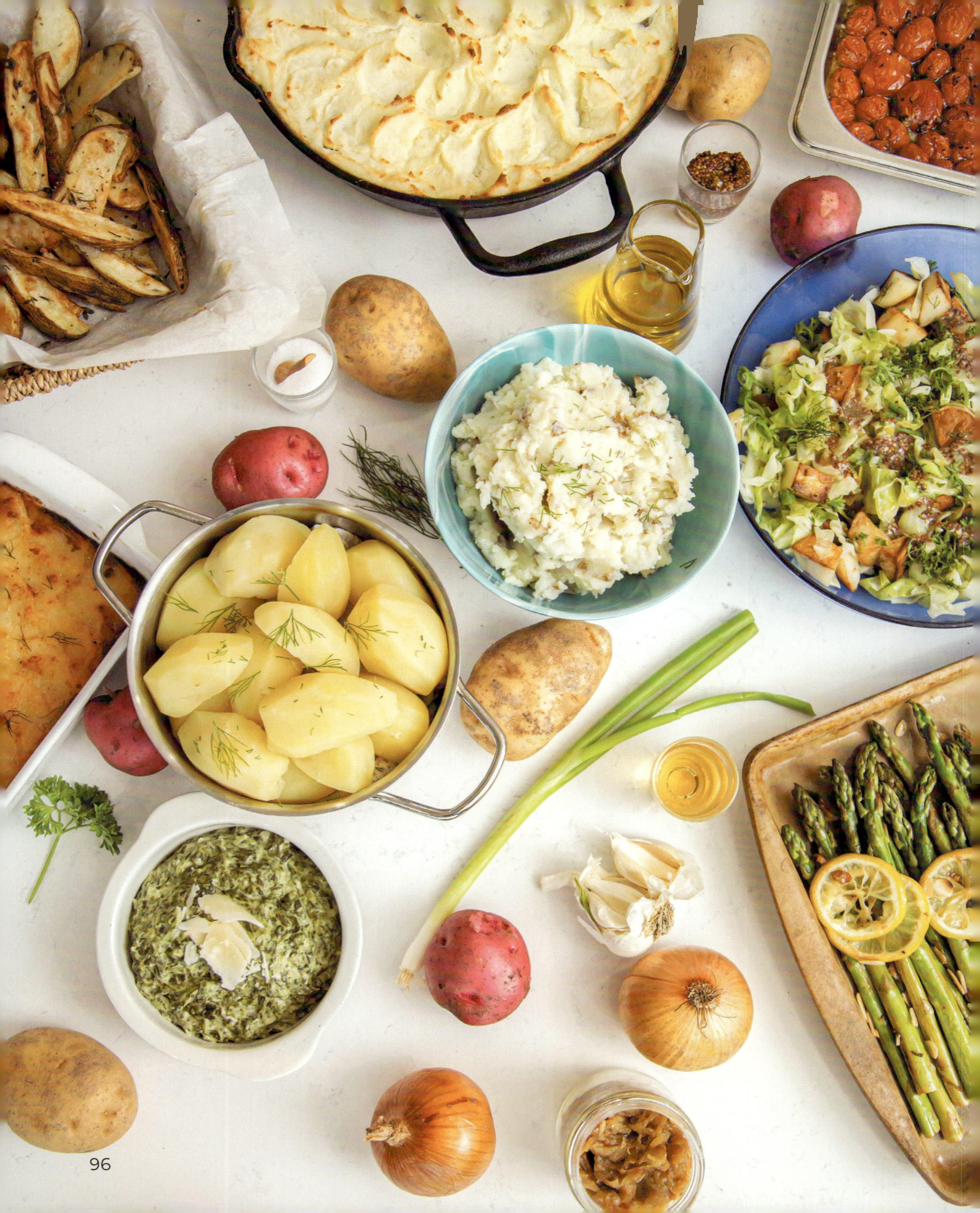

6. Wow Potatoes and Veggie Side Dishes

MY STORY: SIZE FOR SPEED

In our busy household of a growing family of four, where we are off to swimming, gymnastics, or basketball, it is so important to be able to make a wholesome meal fast. And in the fall of the year, when potatoes and onions and other root and cruciferous vegetables are in abundance, I'm looking for a way to dress them up. Potatoes are so versatile and functional that I make them in all kinds of ways. Traditional "boil together in one pot" vegetables are generally overcooked, mushy, flavourless, and boring. I have been jazzing them up over the years by adding seasonings and varying the sizes of the veggies, not just for nicer presentation but also so that heat can distribute better for more uniform cooking and perfect doneness. Since I'm always in a hurry, this method offers faster cooking times as well as better seasoning distribution, so every bite has all the flavours. Potatoes and vegetables in general have delicate flavours, so adding salt, cheese, butter, cream, and herbs make them even more addictive. It all lies in the preparation. Size matters.

THE MAGIC OF SCIENCE: SIZE MATTERS

From a young age, I studied cellular biology in school and I'd look at thin slices of potato under the microscope I had gotten for Christmas. It was fascinating to see the cellular structure of plant foods such as asparagus, broccoli, carrots, and potatoes, to name just a few. Each vegetable has its own composition of starch, moisture, and cellular structure that is exposed and degraded during the cooking process.

Potatoes are typically a lovely blank canvas in that their flavour is simple and mild, so adding salt and seasoning enhances their flavour. During the cooking process, penetrating heat softens the potato starch cell membranes, cooking the potato and activating its ability to hold moisture. The large starch molecule inside the cellular structure unravels, takes on water, and holds it. Making sure all the pieces are cooked at the same time ensures you can have perfect mashed potatoes, steamed carrot coins, or mixed roasted veggies in every bite. How you cut them in the preparation stage is the key to success.

Denser veggies cut in the same dimensions always take longer to soften and cook. Asparagus spears are more delicate than broccoli stalks, so broccoli stalks require more time to cook and soften the cellular structure in the stalk. Cutting the stalks thinly and increasing the surface area for heat penetration encourages cellular breakdown, speeding up cook time. Carrots take twice as long to cook compared to potatoes when cut to the same dimensions. As well, potatoes cut in smaller pieces cook faster than larger pieces of potato due to increased surface area and heat penetration. So, size does matter for penetration of heat and how fast you want your vegetables cooked.

WOW! POTATOES

I'm so impressed when I go to a restaurant and find their diced carrots or wedged potatoes are perfectly sized. They have taken the time to ensure uniformity in size for great visuals and perfect cooking. I bite in, and it is perfection. Every piece is perfectly cooked and uniformly seasoned. This secret is not new. McDonald's fries were carefully designed to take into consideration deep-fry temperature, cook time, and the container they go in, along with a precise amount of salt per deep-fryer basket load.

Let's talk potatoes. You can get various types at the grocery store: white, russet, creamers, red, and Yukon Gold are likely the main ones that grow locally so are most often used. You'll notice potatoes come in all different sizes. You can wash them and cook them with the skins on, or you can peel them to rid them of any exterior colour or tough skin. However, when you cut them, you expose the starches and cut the cells, releasing moisture as well. Depending on the type of potato, the amount of its amylose, amylopectin, and starch molecules differs. Some potatoes react quickly with the open air and turn brown. This is an irreversible reaction. Limiting the exposure to air is usually done by submerging the cut potatoes in water or vinegar. You can also use oil or butter to block air exposure. The medium you use just needs to create a barrier. Wrap the cut potatoes tightly in plastic that is impermeable to oxygen. (Try freezer bags. Most commercial plastic wrap has high rates of oxygen transmission so doesn't work that well.) Less surface area limits the potatoes' exposure to air, so cut them only when you're ready to cook them. And be sure to cut them to equal thickness so all the pieces of potato, whole, halved, quartered, or diced, cook equally and are ready at precisely the same time. Heat, whether it is moist heat (in water, steam, or foil), or dry heat (from roasting or baking) continues to break down cells and degrade the potato. Uneven cooking can create a mixed pot of both hard, undercooked parts and overcooked morsels. Firmer potatoes are fine for making potato salad, but for mashed potatoes you want the morsels to be cooked through.

- Sizing Potatoes
- Boiled Potatoes
- Garlic Smashed Potatoes
- Wedges/Frites
- Duchesse Potatoes
- Shepherd's Pie
- Buttered Potatoes and Cabbage
- Scalloped Potatoes
- Mini Pommes Anna

Rule of thumb: To cook carrots and potatoes at the same time in the same pot and have them be ready to serve at the same time, carrots should be about half the size of the potatoes.

SIZING POTATOES

Peeling: Use a peeler or paring knife to peel potatoes and take off the external skin layer. Being sure to remove the eyes and defects.

Cut: With or without peel

Dice: Large pieces: Cut the potato in 2 cm slices and then again to create 2 cm (1") pieces.

Small pieces: Cut the potato in 1 cm slices and cut again to create 1 cm (½") pieces.

Wedge: Cut the potato in half, then cut in half again on the long edge. Then again on the long edge, cut a 45-degree angle in the middle of that piece.

Chunk: Cut the potato in half and evaluate the size of every other piece, then cut as necessary to achieve consistent thickness.

Rule of thumb: Dice pieces small for fast cooking or soft texture for mashed potatoes.

Dice pieces large for potato salad, stews, and soups to maintain integrity.

Use uniform wedges for even cooking while roasting.

Chunky potatoes are good as a side. They can be large but need to be uniformly sized for equal cooking and penetration of heat.

101

BOILED POTATOES

MAKES 4-6 SERVINGS

Yes, boiling potatoes is thought to be so simple, and even boring. But boiling can create issues for maintaining the integrity of the potato when making potato salad or make the potato pieces too hard to mash (or smash) for smooth mashed or duchesse potatoes.

2 lbs potatoes, uniform size, peel on or off
8 cups water
1 tsp salt

1. Fill a large saucepan with the water (or enough to cover the potatoes completely.) Add the salt to the water. Rinse the excess starch from the raw potatoes.
2. Add the potatoes and distribute them evenly in the pot.

For a side dish or potato salad:	Cook just until a fork easily penetrates the potato.
For mashed or smashed potatoes:	Cook until the potato just breaks apart with the push of a fork.
For duchesse potatoes, potato pancakes, or gnocchi:	Cook until the potato pieces break immediately when you insert a fork.

> **Rule of thumb:** White, red, or Yukon Gold potatoes are best for any dish where the potato needs to hold piece integrity, such as potato salad, a side for entrée, soup, or salad.
>
> Salt will not only enhance the flavour, but it will also increase the boiling point of the water slightly and will break down the potato cells and starches faster leading to creamier mash potatoes.

103

GARLIC SMASHED POTATOES

MAKES 4–6 SERVINGS

4 medium potatoes, washed, unpeeled, and diced large
8 cups water
1 tsp salt

2 cloves garlic, smeared and salted (see Garlic Aioli on p. 48)
½ cup milk
3 tbsp butter
Salt and pepper to taste

1. In a large saucepan, add the water and the cut potatoes. Add salt to the water. Boil until the potatoes break apart with a fork. Strain the water completely.
2. With a masher or fork, roughly mash the potatoes. Add the smeared salted garlic to the hot potatoes and mix vigorously (to cook the garlic). Then add the butter and mix until melted.
3. Add the milk and blend until smooth. Season with salt and pepper. Serve warm.

WEDGES/FRITES

MAKE 3-4 SERVINGS

4 medium potatoes, washed, unpeeled, and cut into wedges
1 tsp salt

½ cup extra virgin olive oil
½ tsp salt
¼ tsp pepper

1. Preheat the oven to 400°F (200°C). Dry the wedged potatoes and add to a bowl.
2. Drizzle the oil all over the potatoes. Add salt and pepper (and other seasonings as desired), and toss until all the wedges are evenly coated.
3. Transfer to a baking sheet lined with parchment paper, making sure the potatoes are in a single layer on the pan. Bake for 20 minutes or until a fork easily pushes into the wedges and the potatoes are a nice golden colour.

FREESTYLE

Add: Herbs
1 tsp dried rosemary
(or any herb of your choice)

Add: A Little Heat
½ tsp chill powder
¼ tsp cayenne powder

Add: Smokiness
½ tsp smoked paprika
½ tsp dried thyme

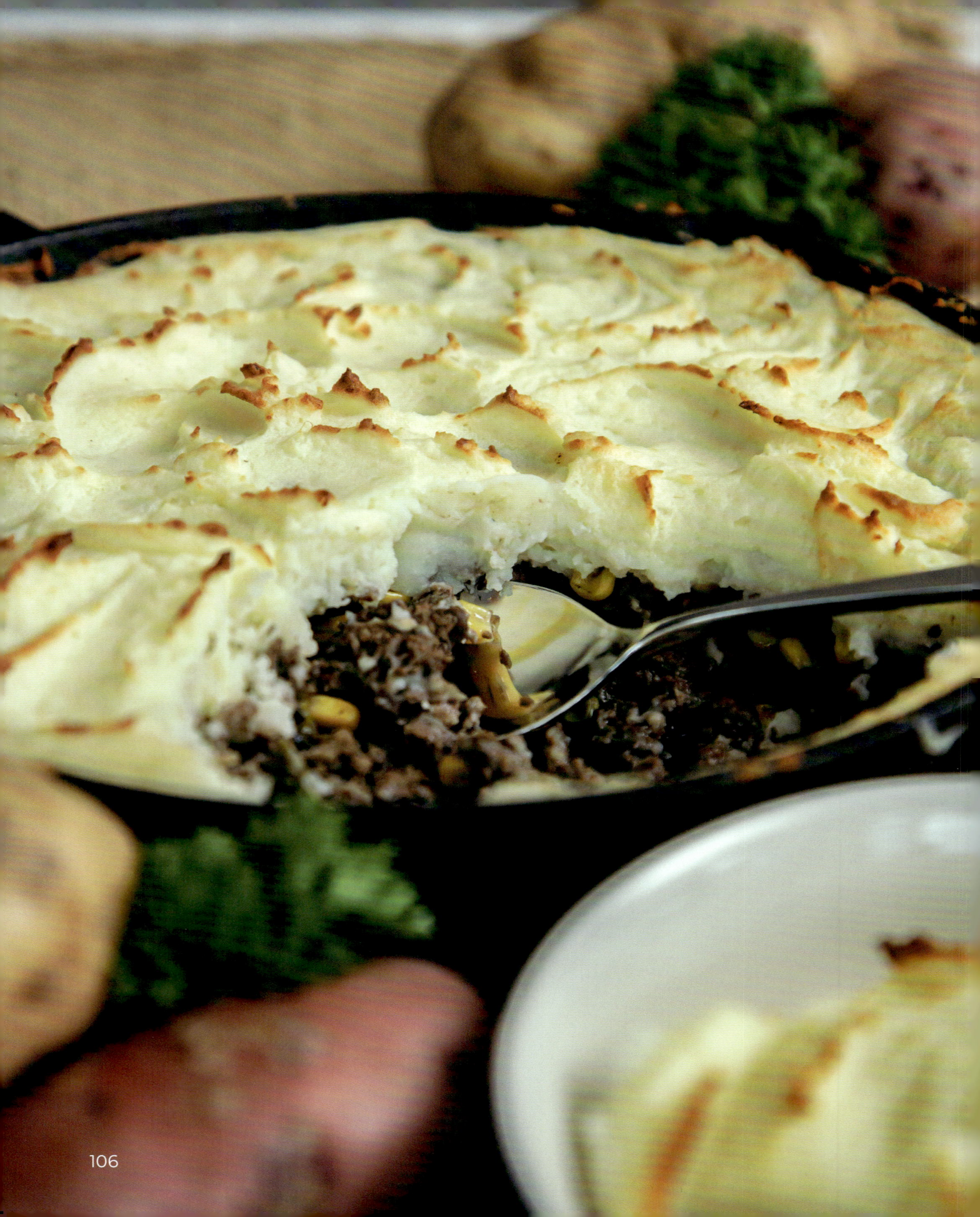

DUCHESSE POTATOES

MAKES 4–6 SERVINGS

4 medium potatoes, washed peeled and cut to large dice (2cm)
1 tsp salt
½ cup milk

1 egg yolk
3 tbsp butter
Salt and pepper to taste

1. In a large saucepan, add 2 L of water and the cut potatoes. Add the salt to the water. Boil until the potatoes break with a fork. Strain off the water.
2. Using the paddle attachment of a standing mixer, mix the potatoes and butter until melted. Add the egg yolk and mix. Add the milk and increase the mixer speed to whip the potato mixture smooth.
3. Season with salt and pepper.
4. Using a piping bag, or tablespoon, portion the potato mixture on top of shepherd's pie or beef bourguignon or just individually on a baking sheet and place under the broiler until the potato peaks are golden brown.

SHEPHERD'S PIE

MAKES 4 SERVINGS

You can substitute ground pork, ground turkey, or ground chicken for ground beef.

1 lb lean ground beef
1 medium onion, chopped finely
2 cloves garlic, minced
1 tbsp extra virgin olive oil
1 tbsp flour
2 tsp thyme

1 tsp Worcestershire sauce
1 cup frozen mixed veggies (peas, carrots, corn, beans)
¾ cup beef broth (or bouillon)
4 servings duchesse potatoes (see above)
3 tbsp butter, cut into small squares

1. Heat the oil in a frying pan. Sauté the onions until translucent. Add the ground beef and cook until the meat is no longer pink.
2. Add the garlic, flour, and thyme. Stir until distributed. Add Worcestershire sauce, broth, and mixed veggies, and simmer until veggies are cooked and the liquid is absorbed.
3. Preheat oven to 375°F. Transfer the meat mixture into a 9" x11" baking dish and spread the meat mixture out evenly over the dish.
4. Using a spoon or a piping bag, portion duchesse potatoes over the meat all the way to the edges. Dot with butter.
5. Bake for 30 minutes until the meat is bubbling and the potato topper shows some golden-brown peaks.

BUTTERED POTATOES AND CABBAGE

MAKES 4 SERVINGS

4 medium potatoes, washed and diced large
1 head savoy cabbage (green cabbage or baby bok choy are good too), roughly chopped
3 tbsp butter
½ cup fresh dill (or 1 tbsp dried)
½ cup fresh parsley (or 1 tbsp dried)
½ cup fresh chives (or 1 tbsp dried)
¼ tsp coarse sea salt
¼ tsp freshly ground black pepper
3 tbsp cider vinegar or white wine vinegar
1 tbsp grainy Dijon mustard
3 tbsp vegetable oil

1. In a large frying pan, add 1 tbsp butter and the potatoes. On medium heat, cook the potatoes for 5 minutes without stirring until the bottoms of the potatoes are brown, then turn the pieces to brown the other side. Transfer to a serving dish and keep warm.
2. Add 2 tbsp butter to the frying pan. Melt the butter and add the cabbage. Cook and stir until all the cabbage pieces are bright green and tender and soft. Transfer to the serving dish on top of the potatoes.
3. Sprinkle the herbs, salt, and pepper over the cabbage.
4. Mix the vinegar, oil, and mustard together, and drizzle over the entire dish.

109

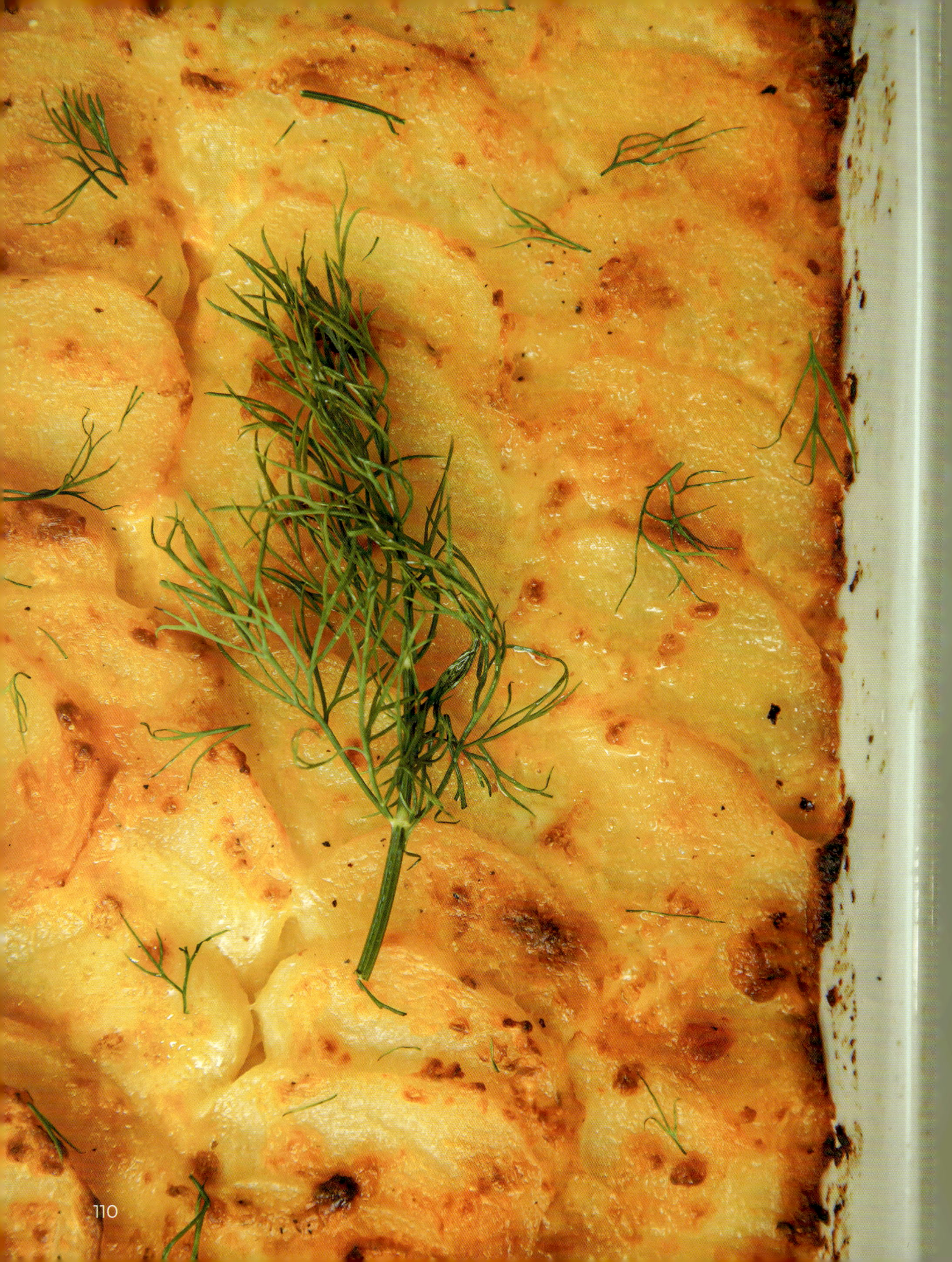

SCALLOPED POTATOES

MAKES 4–6 SERVINGS

4 medium potatoes, sliced very thin and submerged in water
2 cups thin béchamel sauce (see p. 84)
2 cups old cheddar cheese, sliced thin

2 tbsp butter
1 large onion, thinly sliced
Salt and pepper to taste

1. Preheat the oven to 375°F.
2. Grease a 9"x11" baking dish with cooking spray oil.
3. Place the sliced potatoes in the bottom of the dish in a single layer, overlapping the potato slice (similar to shingles on the roof of a house). Add a layer of onions, salt, and pepper, then add cheese.
4. Cover with enough of the béchamel sauce to cover the cheese-onion-potato layer.
5. Repeat steps 3 and 4 until all the potatoes are in the dish or the dish is full. Top with the remaining cheese.
6. Place a baking sheet under the dish and bake for 45–60 minutes until the sauce is bubbly and the potatoes are cooked.

FREESTYLE

Substitute béchamel sauce with 1 cup milk. Pour the milk over the entire dish once the potatoes, cheese, and onions are all arranged.

Substitute béchamel sauce and old cheddar cheese with premade thin Mornay sauce (see p. 86).

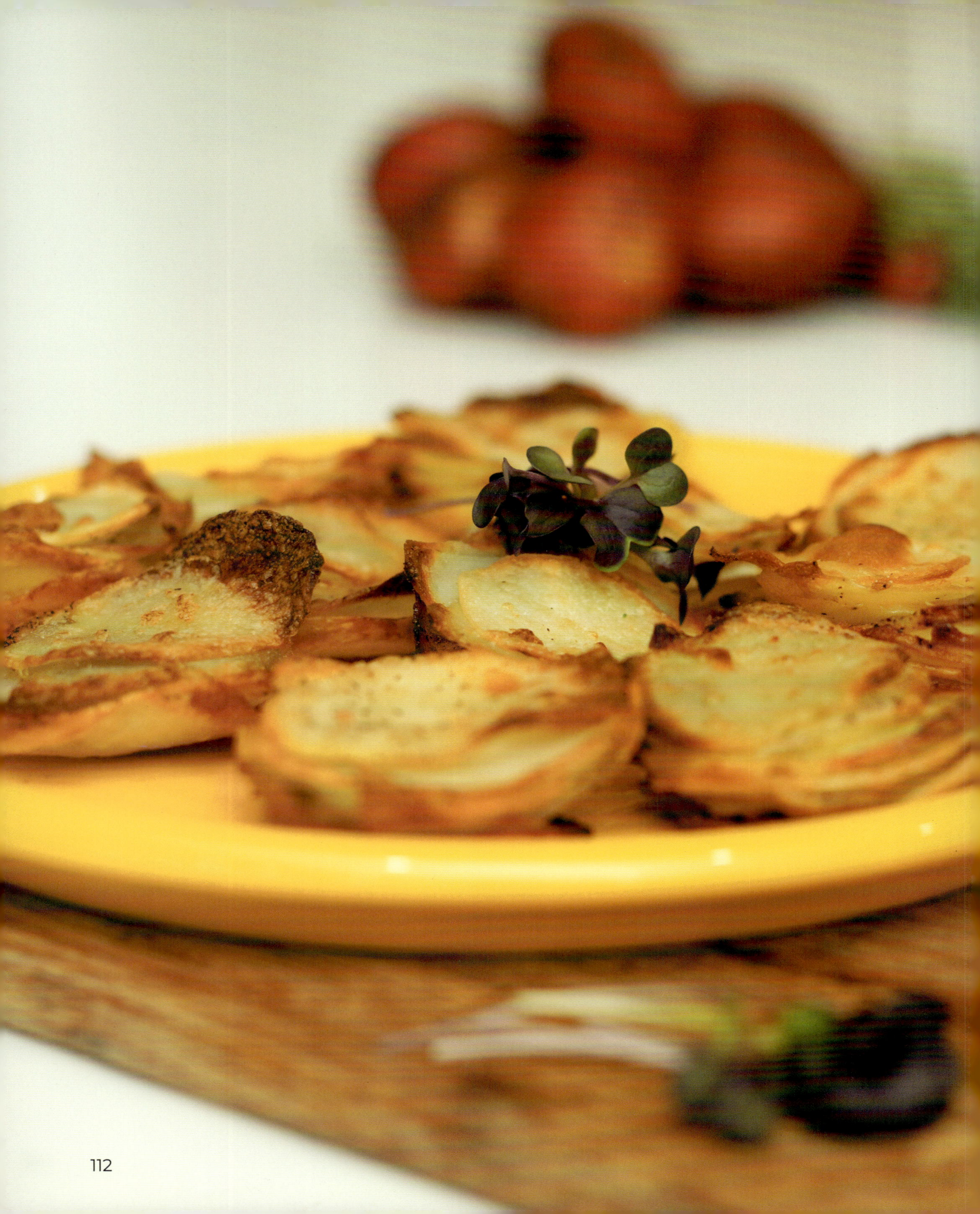

MINI POMMES ANNA

MAKES 4 SERVINGS

4 medium potatoes, sliced very thin
⅔ cup butter plus 2 tbsp for the muffin pan
2 tbsp butter

2 cloves garlic, minced
Salt and pepper to taste

1. Preheat the oven to 375°F. Prepare a muffin pan by distributing the 2 tbsp of butter equally among the 12 muffin cavities.
2. Place the muffin pan in the oven for a few minutes to melt the butter.
3. Melt 2/3 cup of butter in a large saucepan so it is just warmed. Add the garlic to it to infuse for a moment.
4. Add the thinly sliced potatoes to the saucepan with the warmed garlic butter. **Be sure to cover the potatoes with the garlic butter so they don't change colour.**
5. Slice by slice, arrange the buttered potato slices in the muffin cups and organize them so they are layered to the rim. After every third or fourth slice, season with salt and pepper.
6. Bake for 45 minutes until crispy on the edges and on top.
7. Let cool for 10 minutes before taking them out of the muffin pan.

Can be made ahead, stored in the refrigerator, and reheated on a baking sheet for 10 minutes at 400°F.

VEGGIE SIDE DISHES

Potatoes or some other starch always seems a given for dinner and an accompaniment to a main entrée protein. Jazzing up other vegetables is just as easy as jazzing up potatoes. Here are a few of my favourites that can be prepared very quickly, add more nutrition, flavour, and colour to your meal.

- Creamed Spinach
- Tomato Confit
- Caramelized Onions (Confit d'Oignon)
- Asparagus, Lemons, and Almonds
- Roasted Vegetable Medley: Mixed Vegetables
- Roasted Vegetable Medley: Broccoli, Brussels Sprouts, and Cabbage Veggie Bake

CREAMED SPINACH

MAKES 4 SERVINGS

3 tbsp butter
2 tbsp flour
Pinch nutmeg
½ cup heavy cream
1 clove garlic
¼ cup cream cheese
Salt and pepper
½ medium onion, finely chopped
1 ¼ cups frozen spinach (or 5 cups fresh and washed)
¼ cup parmesan cheese

1. Thaw the spinach and drain off the excess moisture.
2. In a large frying pan, sauté the onion in the butter until soft and translucent.
3. Add the garlic and cook for an additional minute.
4. Stir in the flour and cook for 1–2 minutes.
5. Add the cream and cream cheese and stir. The mixture should be thick.
6. Add the spinach until it is warmed.
7. Season with parmesan cheese, nutmeg, and salt and pepper.

TOMATO CONFIT

MAKES ABOUT 1 CUP

1 lb grape or cherry tomatoes
2 cloves garlic, minced
¼ cup olive oil

2 tbsp lemon juice
1 tbsp honey
1 tbsp basil, freshly chopped

1. Preheat the oven to 275°F.
2. Pour the oil over the tomatoes, add the herbs and garlic cloves, and season generously with salt and pepper. Toss to combine.
3. Put the tomatoes on a baking pan large enough to hold them in a single layer. There is no need to grease the pan or use parchment paper since the tomatoes have oil on them and the juice will soak the parchment.
4. Roast the tomatoes, uncovered, until the tomatoes are swollen, and the skins are wrinkled, about 2 hours.
5. Remove the tomatoes from the oven, allow to cool, then pour the tomatoes into a small bowl. Refrigerate overnight to develop the flavours.
6. When you are ready to serve, mix gently to combine the liquid and tomatoes, being careful not to break up the tomatoes.

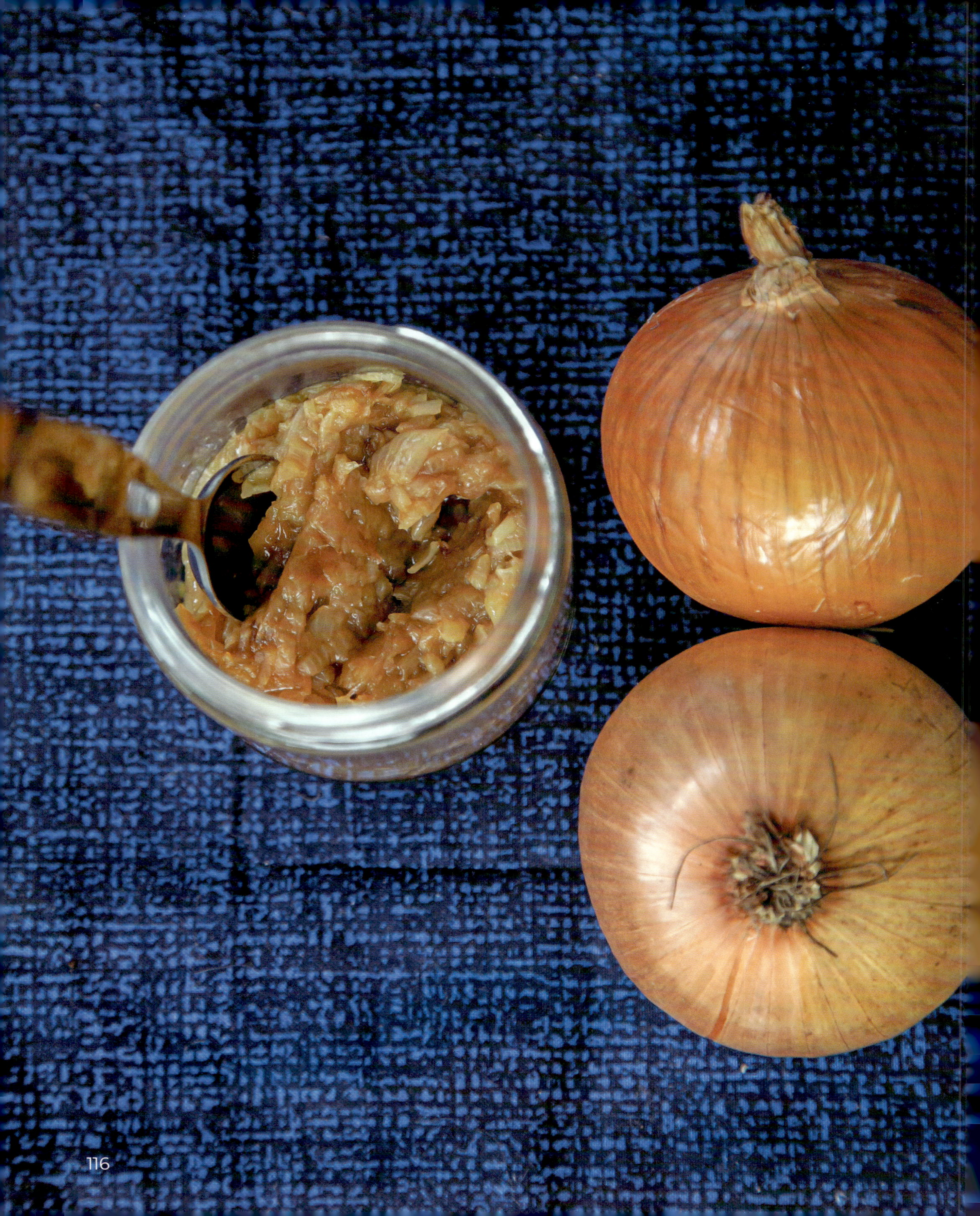

CARMELIZED ONIONS (CONFIT D'OIGNON)

MAKES ABOUT 1 CUP

4–6 medium onions
1 tsp salt

1 tbsp butter
Pepper and nutmeg, as desired

1. Slice the onions into 3-mm thick slices and place them and the butter in a large frying pan that has a lid. Add the salt and cook the onions over high heat until they begin to sizzle, then lower the heat to the lowest possible setting.
2. Tightly cover the onions with a lid and allow to simmer them in their juices for 1–3 hours. The longer they simmer in their juices, the better the flavour.
3. When ready to serve the onions, increase the heat to evaporate the juices and develop a golden colour. Transfer to a bowl and cool.
4. Before reducing the juices, spoon some out to moisten roasted veggies or meat.
5. Use the finished confit when plating or as a flavour-enhancing addition for simmering sauces or dishes of sautéed potatoes. This confit can take the place of bouillon.

FREESTYLE

You can puree the mixture before all the juices are reduced and the onions are caramelized. Add fresh herbs and a little cream to create a herb-infused onion sauce.

ASPARAGUS, LEMON, AND ALMONDS

MAKES 4 SERVINGS

12 asparagus spears, trimmed
1 garlic clove, chopped fine (or 1 tsp minced)
1 tsp lemon juice
½ tsp grated lemon zest
1 tbsp olive oil
½ tsp salt
¼ tsp pepper
¼ cup cooked bacon, chopped (optional)

1. Preheat the oven to 400°F.
2. Toast the almonds in a dry frying pan.
3. On a baking sheet, lightly dress the asparagus and garlic with olive oil, lemon juice, and zest and season with salt and pepper.
4. Roast the asparagus in the oven for 18–20 minutes until the asparagus is tender and slightly charred in places.
5. Transfer to a plate and top with almonds and bacon bits.

HOW TO TOAST ALMONDS AND OTHER NUTS

Place the sliced almonds in a cold frying pan. Turn the stove on medium low. Let the pan heat up and warm the almonds. Once they are toasted on one side, toss the almonds to toast the other side. Once they look toasted, quickly remove the almonds from the heat and pour them into a shallow dish to cool. Do not leave the almonds in the frying pan as they will continue to heat and will burn quickly.

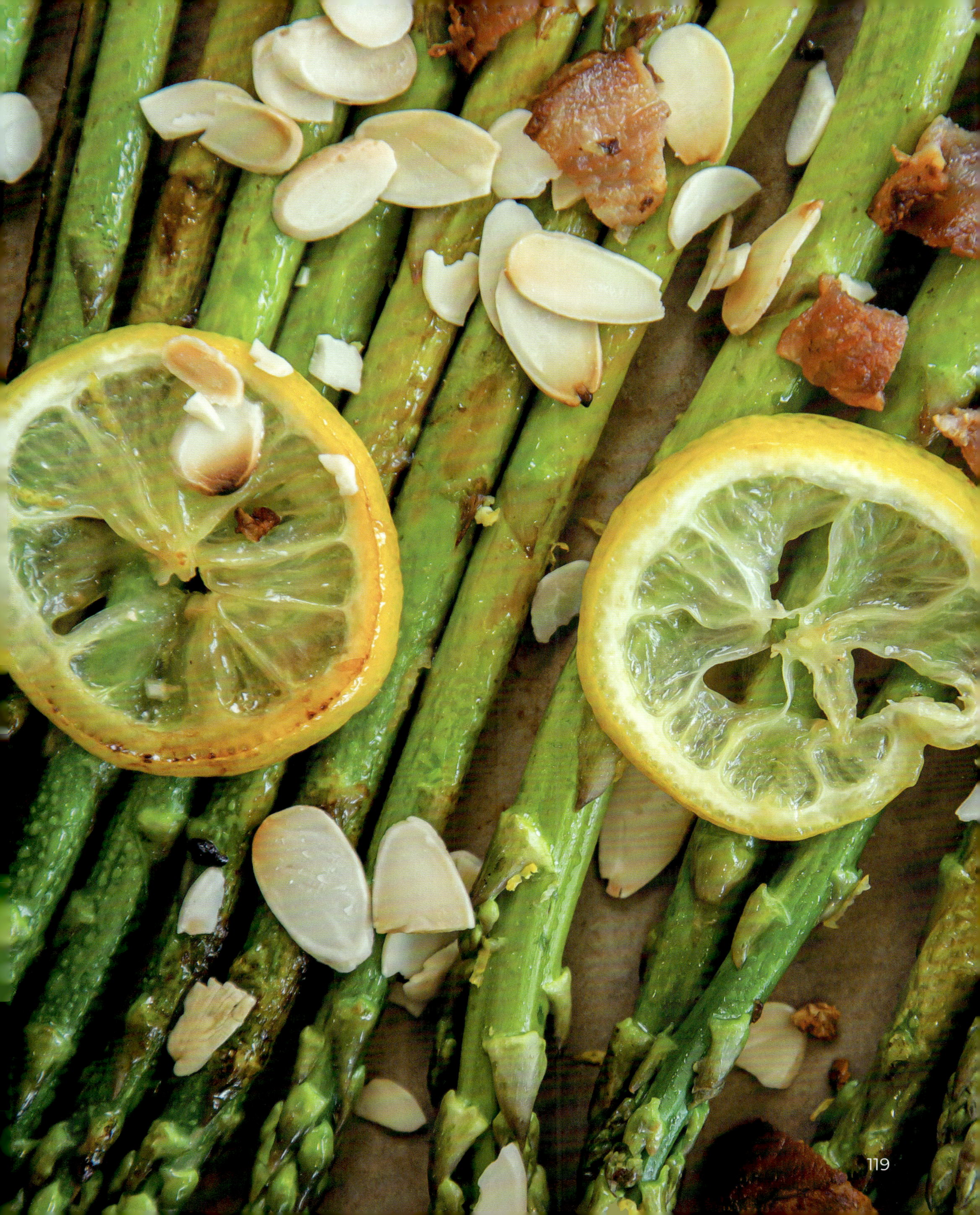

ROASTED VEGETABLE MEDLEY: MIXED VEGETABLES

MAKES 4 SERVINGS

1 cup red peppers, cut in 1" pieces
1 cup green peppers, cut in 1" pieces
1 cup zucchini, cut in 1" chunks
1 cup onion, peeled and cut in 1" chunks
½ cup extra-virgin olive oil

3 cloves garlic
½ tsp salt
¼ tsp pepper
1 teaspoon Italian seasoning

1. Preheat the oven to 375°F. Line a baking sheet with parchment.
2. Combine the vegetables in a bowl.
3. In a separate small bowl, blend the oil, garlic, Italian seasoning, salt, and pepper.
4. Pour the oil mixture over the veggies and combine well.
5. Spread the mixture evenly on the baking sheet. Bake for approximately 25 minutes, until the peppers are a little charred and the onion is soft.

ROASTED VEGETABLE MEDLEY: BROCCOLI, BRUSSELS SPROUTS, AND CABBAGE VEGGIE BAKE

MAKES 4 SERVINGS

1 cup kale, chopped
1 cup broccoli florets
1 cup Brussels sprouts, cut in half
1 cup cabbage, chopped
½ cup butter, melted

3 cloves garlic
½ tsp salt
1 tsp allspice
½ cup slivered almonds

1. Preheat the oven to 375°F
2. Combine all the vegetables in a bowl.
3. In a separate small bowl, blend butter, garlic, salt, and allspice. Pour over the veggies and combine well. Add the dressed veggies into an ungreased 9"x13" baking dish.
4. Sprinkle almonds over the mixture
5. Bake for 45 minutes until the vegetables are soft and browned. Eat hot or cold.

122

7. Get Crackin' with Eggs

MY STORY: FLAWLESS CRÈME BRÛLÉE

One winter our family headed to a remote resort in Quebec for a holiday of snowmobiling. While the trails for snowmobiling were outstanding, it was the food in the evenings we looked forward to. We always chose a decadent poached-apple crème brûlée, topped with a little snowman made of dehydrated apple slices on a stick. Curious, I proceeded to deconstruct the dessert and figured out how I could make it at home. The crème brûlée I made was tasty but never had a crème texture. I went back to basics and tried different ratios of sugar, eggs, and cream, but the dessert never turned out. I wasted many eggs as well as lots of cream and was embarrassed each time. I'd declare it should work, and it didn't. It was not until I was shown what I needed to do that I achieved success. I learned it's all in how the functional components of the egg are activated. Essentially, I was not whisking the egg yolks long enough to unravel the proteins so they could work to create a gel matrix. As well, at high temperatures, the eggs would cook too quickly, and the crème would curdle. Fixing the preparation and cooking process eliminated those embarrassing crème brûlée failures. I now make flawless crème brûlée of all varieties, and you can too.

THE MAGIC OF SCIENCE: EGG GEL NETWORKS

What is it about eggs? They are highly nutritious and easily digestible. Their structure explains their amazing functionality. The whole egg contains complex molecules of protein that lie in coils. When egg is whisked vigorously, the protein coils unravel or unwind into long strands. These protein strands incorporate air and form a network of bubbles that stabilizes foam. Since whole eggs contain both the yolk and the albumen, they can be beaten for a long time, but if just the egg whites are over whipped, the protein coagulates so much the egg whites start to separate. When an egg is heated, its protein converts into a network of strands that tighten during cooking, creating a gel with elastic properties that give the ability to create texture (thick or thin) and structure (firm or smooth). If overcooked, these tightened strands of protein shrink, pushing the air and moisture out, making the egg tough and yielding less volume. Egg whites start to create this gel at 145°F (62°C) and egg yolks at 155°F (70°C) or slightly higher when diluted with liquid.

On ne fait pas d'omelette sans casser des œufs.
-French proverb

THE FUNCTIONALITY OF EGGS

Eggs can accomplish greatness or create awful flops. There are only three things to remember in working with eggs. That's it!

- The right amount of mechanical action to expose the aeration and gelling properties of the proteins and the emulsification properties of the fats
- The right amount of time
- The right temperature

With egg yolks, it's so important to whisk, stir, or agitate them until you see a colour change. This is the indicator that lets you know the proteins have been unraveled and aerated to avail its transformational properties. Thanks to the presence of two emulsifying agents in egg yolks, cholesterol and lecithin, fat and water molecules combine to ensure success in products such as custards, sauces, cakes, mayonnaise, hollandaise, and even cookies and some pastry.

For egg whites, whisk and aerate the protein albumen, but don't over whip. Egg whites will extend the elasticity properties for each dish you make. For our granola, that elastic property is what helps to keep the granola clusters together.

- Fluffy Minute Scrambled Eggs
- Boiled Eggs
- Deviled Eggs
- Spinach and Bacon Breakfast Frittata
- Mustard Baked Eggs
- Hollandaise Sauce
- Béarnaise Sauce
- Butter-Poached Lobster Eggs Benedict
- Chocolate Éclairs
- Lemony Lemon Meringue Pie
- Chocolatey Chocolate Mousse
- Flawless Crème Brûlée
- Granola
- Pavlova

Rule of thumb: Use fresh eggs – as fresh as possible. As the egg ages, the enzymatic activities continue and break down the functional properties of the protein, particularly in egg whites, and release moisture.

If you use older eggs, your meringue will be flat, the frittata will be less fluffy, poached eggs will be mostly yolk, and chocolate mousse will be dense.

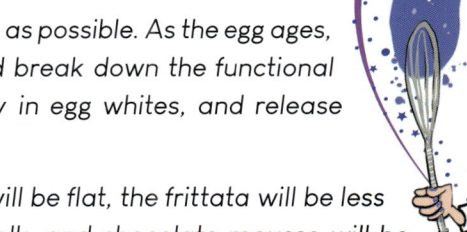

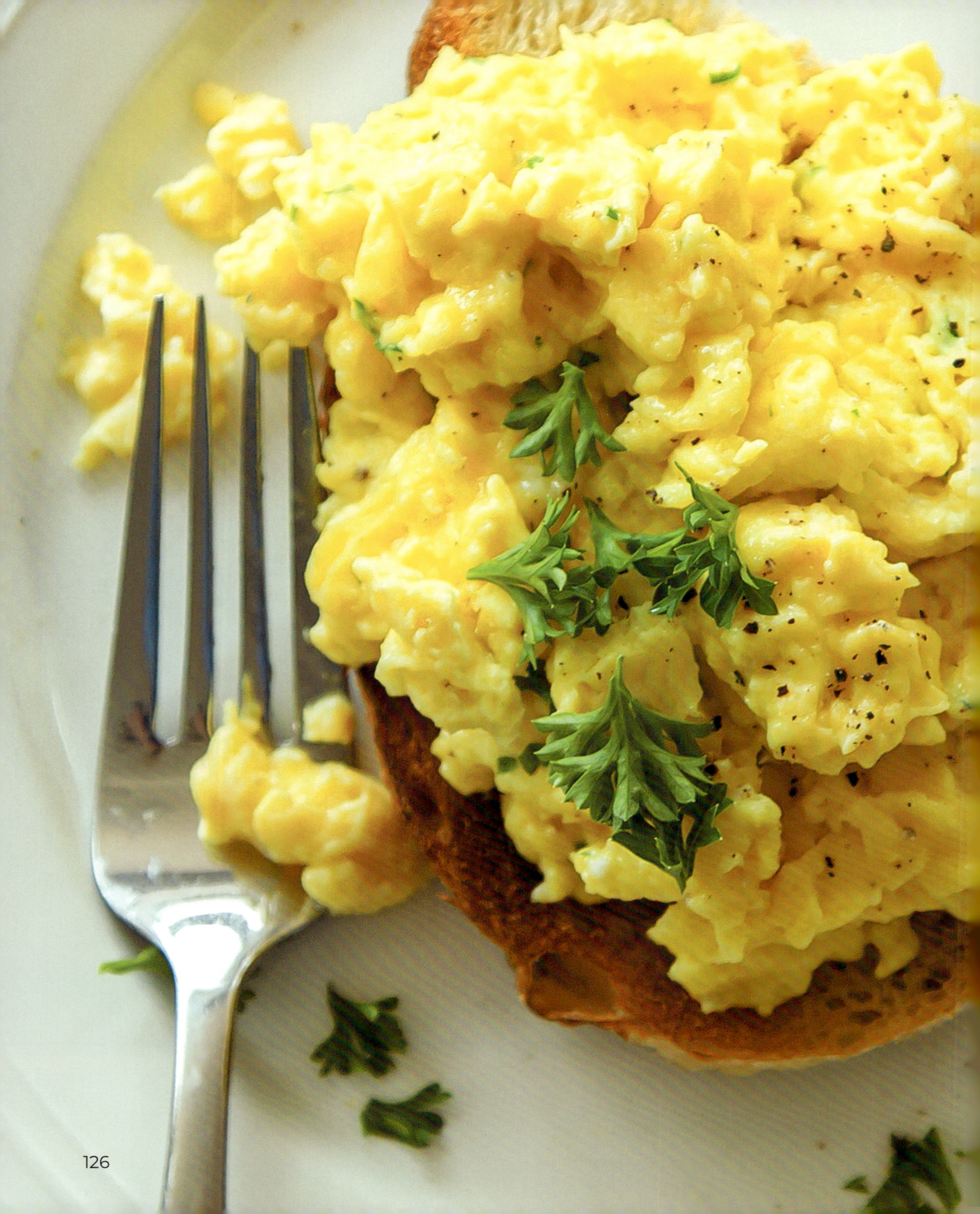

FLUFFY MINUTE SCRAMBLED EGGS

MAKES 1 SERVING

2 large eggs
Dash of salt
Dash of pepper

1 tsp parsley, finely chopped
1 tbsp butter

1. Break the eggs into a bowl. Whisk for 1 minute until they are combined and frothy.
2. Add the dash of salt and pepper and parsley just before cooking.
3. Heat a small frying pan on high; melt the butter until it sizzles. Turn the heat to medium.
4. Add the egg mixture to the pan. The egg that comes in contact with the pan cooks first, so slowly move the eggs around the pan with a wooden spoon, continuing the process until the creamy curds form.
5. Take the pan from the heat and stir a little more. The latent heat from the pan continues to cook the eggs for a minute until they are sufficiently thickened.
6. The eggs should be light yellow and fluffy and in ample quantity for a generous serving.
7. Tip onto your plate, taste, and season as needed.

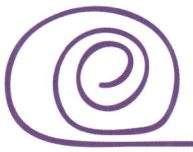

Rule of thumb: Watch the heat!

Aggressive or prolonged cooking results in less egg volume to eat. High heat causes the proteins to tighten and shrink, pushing moisture out and typically causing some unintentional browning (Maillard reaction). We often refer to this as curdling.

Stirring the eggs distributes them to the heat and cools the pan with air velocity.

A good nonstick pan enables quick movement around the pan for equal heat distribution (see Get Ready! on p. 6).

For more on scrambled eggs, check out the "How To" video on YouTube.

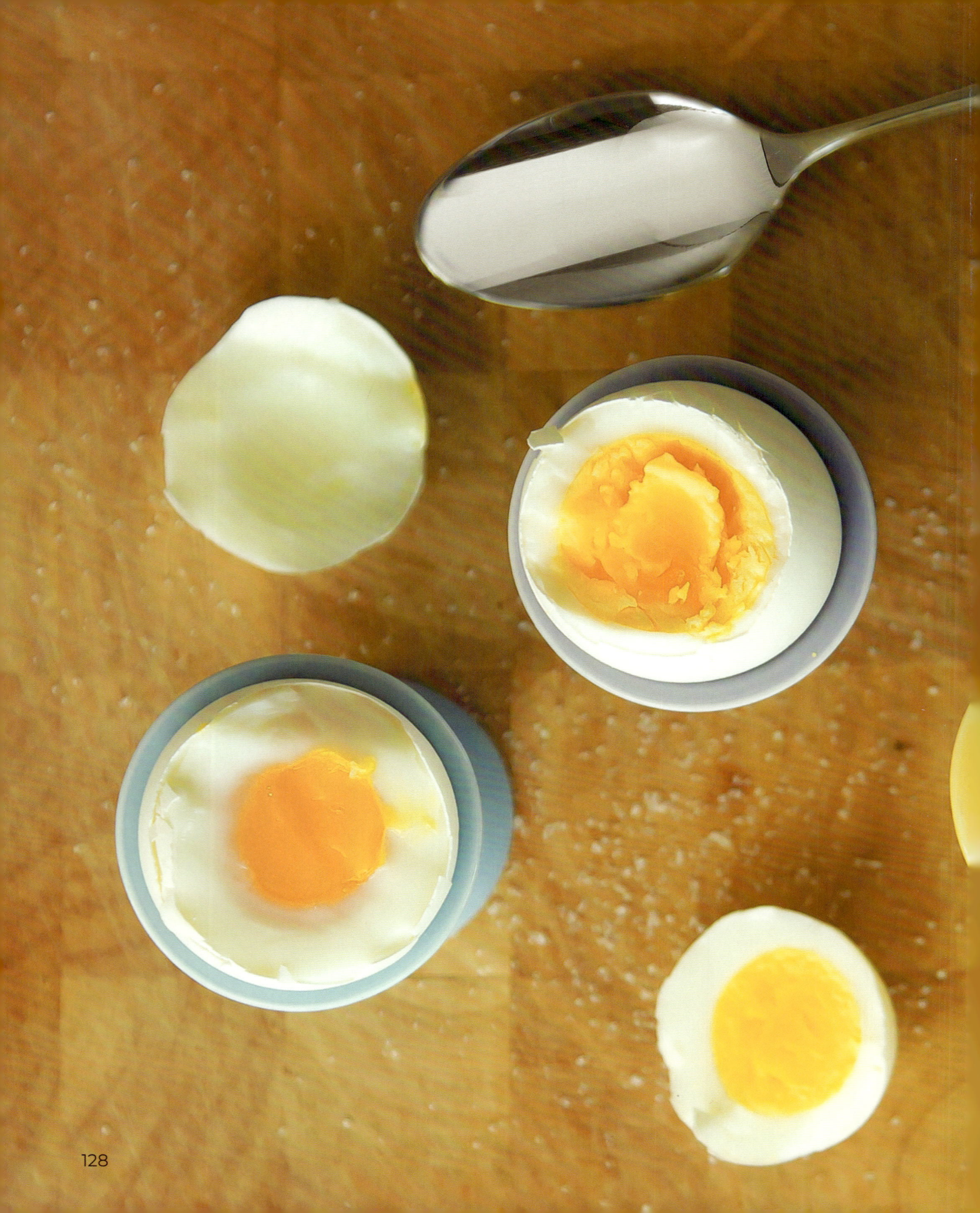

BOILED EGGS
MAKES 1 SERVING

2 large eggs
1 tsp salt
8 cups water

1. Add the eggs to a saucepan. Add water enough to cover the eggs completely.
2. On high heat, bring the water to a boil. Let the water boil gently for 2 minutes, then turn the heat down to low.
   - For soft eggs, allow them to cook for another 6 minutes.
   - For medium eggs, allow them to cook for another 7 minutes.
   - For hard eggs, allow them to cook for 8 minutes.

Rule of thumb: To peel eggs for deviled eggs or a Cobb salad (pp. 130 & 57), run the hot cooked eggs under cold water. The excessive heat causes the steam to be trapped inside the layer between the cooked egg and the shell, creating a space to easily lift off the shell.

Peel the eggs while warm or chill them quickly. If the unpeeled egg chills completely, the moisture reabsorbs, and the shell is more difficult to remove.

DEVILED EGGS

MAKES 4 SERVINGS

4 hard-boiled eggs, peeled
¼ cup mayonnaise
1 tsp Dijon mustard

Juice from ½ lemon
Salt and pepper to taste
Paprika to taste

1. Cut the eggs in half lengthways. Scoop out the yolk and put all the yolks in a bowl together.
2. Add the mayonnaise, mustard, and lemon juice to the egg yolks. Mix everything until smooth.
3. Spoon the yolk mixture back into the egg whites or add the mixture to a piping bag with a star tip and fill the egg whites. Sprinkle with a little paprika or garnish with microgreens.

SPINACH AND BACON BREAKFAST FRITTATA

MAKES 4 SERVINGS

8 large eggs
½ cup milk
3 tbsp extra-virgin olive oil
½ cup onions, diced small
¼ cup cooked bacon, chopped
1 clove garlic, finely chopped (½ tsp minced)
¼ cup potatoes, cooked and diced
2 cups fresh spinach
(or ½ cup cooked), chopped
¼ cup parmesan cheese
¼ cup cheddar cheese, grated
Pinch chili pepper flakes
¾ tsp salt
¼ tsp pepper

1. Preheat the oven to 350°F.
2. Heat the olive oil in a 10" oven-safe frying pan over medium-high heat.
3. Add the diced onions. Cook, stirring occasionally, until softened, about 5 minutes.
4. Add the bacon, garlic, potatoes, and fresh spinach into the pan. Stir until heated through and the spinach is wilted.
5. In a medium bowl, vigorously whisk the eggs for at least 2 minutes.
6. Add the milk, salt, and pepper.
7. Pour the eggs into the pan, stir, and cook gently just until the edges start to pull away from the pan, 5–7 minutes.
8. Sprinkle the chili flakes, parmesan, and cheddar cheese over the top of the mixture.
9. Transfer the frying pan to the oven and bake until set, 16–18 minutes.

MUSTARD BAKED EGGS

MAKES 4 SERVINGS

2 tsp powdered mustard
2 cups sharp cheddar, grated
8 eggs
8 tbsp heavy cream
4 tbsp milk
1 tsp salt
⅛ tsp cayenne
4 tbsp butter or margarine

1. Combine the mustard with 1 tsp water. Let stand for 10 minutes to let the flavour develop.
2. Sprinkle the cheese into a 9x11 baking dish or into 4 individual dishes
3. Break the eggs over the cheese, being careful not to break the egg yolks (two eggs per serving).
4. Combine the mustard, cream, salt, and cayenne and pour over the eggs. Dot the eggs with the butter.
5. Bake at 350°F for 20–25 minutes or until the eggs are set and the cheese is melted.
6. Serve immediately, sprinkled with paprika or chives.

HOLLANDAISE SAUCE

MAKES 1 CUP

¾ cup butter
3 tbsp water
3 egg yolks
1 tbsp lemon juice
Salt and pepper to taste

1. Put the egg yolks in a bowl or the top of a double boiler and whisk them a *lot* "to the ribbon." Stop whisking when the egg yolks change colour from bright yellow to light yellow.
2. Add the water and the lemon juice and mix.
3. In a small pot, melt the butter until just melted and you can still touch it. If the butter is too hot, allow it to cool.
4. Heat water in the bottom of a double boiler and turn down to a slight simmer.
5. Put the top section of the double boiler with the beaten egg yolk mixture on top of the double boiler bottom to warm the eggs just to the touch (no hotter).
6. Add the butter to the eggs a tablespoon at a time for the first 6 tbsps while whisking to incorporate, then stream the rest of the butter into the eggs.
7. Keep whisking until the mixture gets thick. If you need to add a little heat on the water, do so gently, turning it up a little at a time to prevent the egg mixture from curdling.

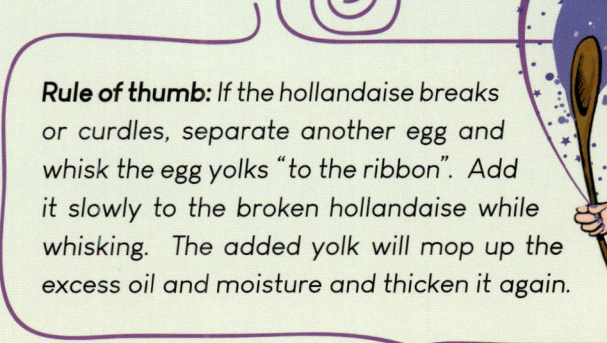

Rule of thumb: If the hollandaise breaks or curdles, separate another egg and whisk the egg yolks "to the ribbon". Add it slowly to the broken hollandaise while whisking. The added yolk will mop up the excess oil and moisture and thicken it again.

BÉARNAISE SAUCE

MAKES 1 CUP

¾ cup butter
3 tbsp white wine
3 tbsp white wine vinegar
3 shallots, finely chopped

10 peppercorns, crushed
1 tsp dried tarragon
3 egg yolks

1. Sauté the shallots in a tbsp of butter; add the white wine, vinegar, and tarragon, then reduce the volume of the mixture to about 1 tablespoon by simmering it over medium heat.
2. Add a tbsp of water to cool the mixture.
3. Prepare the egg yolks in the same way as the hollandaise sauce.
4. Warm over the heat for about 4 minutes.
5. Slowly add in the butter just as you do with hollandaise sauce. As the sauce thickens, add in the tarragon, shallots, and wine mixture.
6. Serve warm with steak, game, or salmon. So good!

135

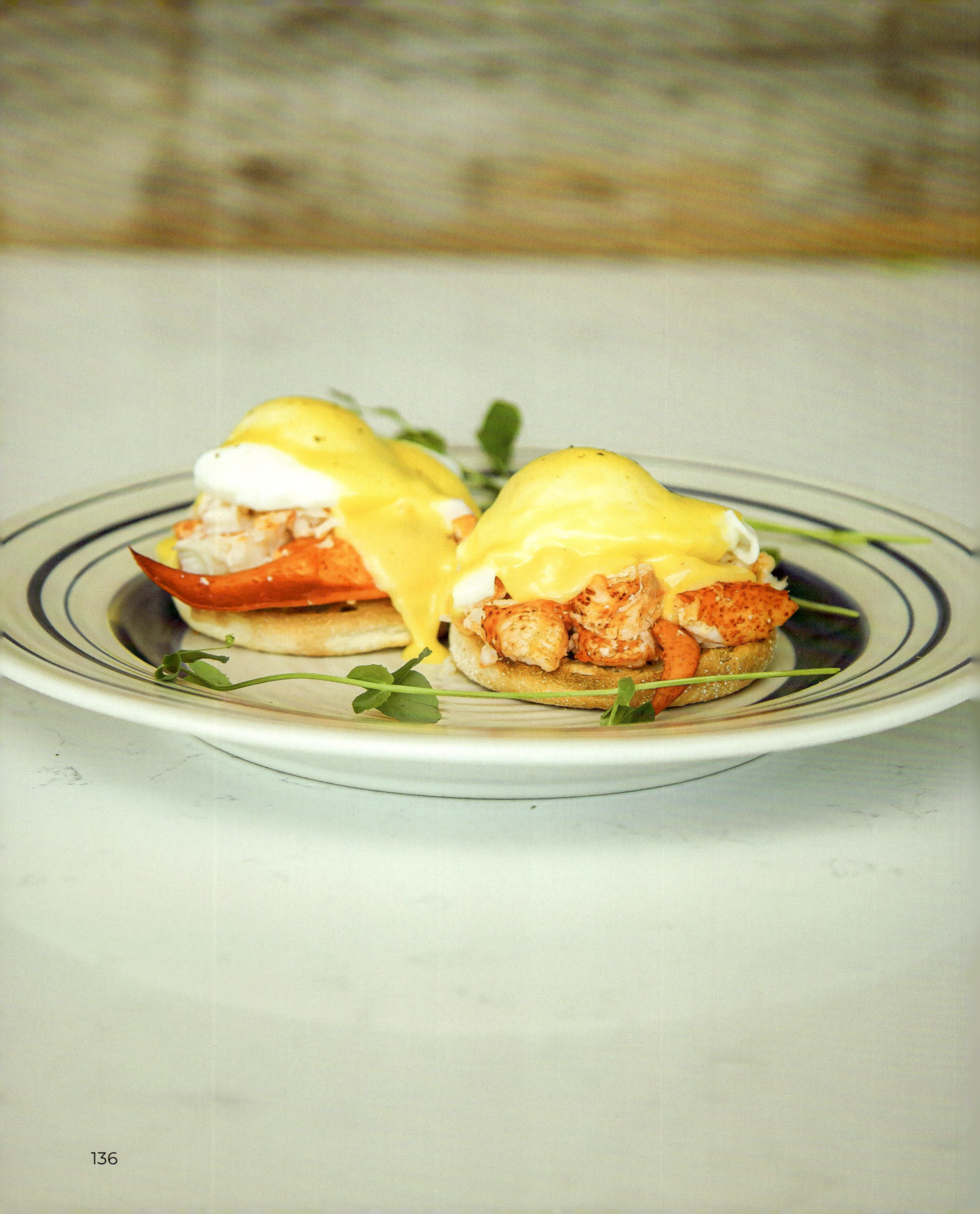

BUTTER-POACHED LOBSTER EGGS BENEDICT

MAKES 4 SERVINGS

Imagine lobster as an option for brunch! It goes so well with eggs and hollandaise. This is our signature dish at Java Jack's, and it's a crowd favourite. Wait until you try it yourself.

8 eggs
4 English muffins
2 cooked lobsters

¼ cup butter
1 cup hollandaise sauce (see p. 133)
Red roe (optional)

1. Poach the eggs just until they're soft. Keep them in warm tap water until you're ready to assemble each serving.
2. Shuck the lobster claws, knuckles, and tails. Cut into ½" (1cm) cubes. Melt the butter in the frying pan and poach the lobster, stirring enough to ensure the lobster is heated through.
3. Make the hollandaise sauce.
4. Toast the English muffins until golden. There is no need to butter the English muffins because you will add the butter-poached lobster and then the soft egg. Top with 2 tbsp of the hollandaise per serving. Crumble some red roe and sprinkle on top as a garnish.

Laughter is brightest where food is best.
-Irish proverb

CHOCOLATE ÉCLAIRS

MAKES ABOUT 12 ÉCLAIRS

Making éclairs takes a little extra time but the result is well worth it. They look special and taste delicious. Éclairs have three components: a traditional choux pastry creates the shell, pastry cream goes inside the shell, and the filled shells are dipped in chocolate glaze.

CHOUX PASTRY

1 cup water
1 tsp salt
1 tsp sugar

½ cup butter
1 cup flour
3 eggs

1. Preheat the oven to 425°F. Line a baking sheet with parchment paper. Prepare a pastry bag with a large plain or star tip.
2. Place the water, salt, sugar, and butter in a saucepan. Bring to a boil.
3. Remove from the heat and add all the flour. Beat the dough vigorously by hand.
4. Put the mixture back on the stove to heat, and continue beating the mixture until the dough comes away from the sides of the pan.
5. Transfer the dough to a stand mixer fitted with a paddle attachment. (You can use a wooden spoon here and work the dough by hand if you don't have a mixer.) Then add the eggs one at a time, beating until the mixture is shiny but firm. The dough should pull away from the sides of the bowl in thick threads. Use a rubber spatula to take this gooey mixture out of the mixing bowl.
6. Put a workable amount of dough in the pastry bag and pipe the dough on the parchment in strips about 1"(2.5 cm) wide and 2.5"(6 cm) long to create the éclair shells.
7. Bake immediately at 425°F for 10 minutes, then reduce the heat to 375°F and bake another 10 minutes. Continue to gradually reduce the oven temperature every few minutes until you reach 200°F or until the éclair shells are brown, dry, and hollow inside.
8. Cool completely.

PASTRY CREAM

3 tbsp flour
1 tsp salt
⅔ cup sugar

1 ¾ cup milk
6 egg yolks
2 tsp vanilla

1. Whisk the egg yolks "to the ribbon" (i.e., whisk until the egg yolks turn light yellow and the consistency is thick enough to create a visible swirl that resembles a ribbon).
2. Add the sugar and mix until the sugar is almost dissolved.
3. Add the flour, salt, and vanilla. Mix until combined.
4. Warm the milk to lukewarm – about 20°C. Add the warm milk to the egg mixture. Whisk to combine.

5. Turn up the heat to medium and cook gently until the cream begins to bubble. Allow a few big bubbles to break the surface before you chill the cream.
6. Chill the pastry cream on a bowl of ice or in the snow so the bottom of the bowl does not continue to heat and burn the cream that is close to the bottom.

CHOCOLATE GLAZE
1 cup (250 g) dark or semisweet chocolate
½ cup butter
1 tbsp corn syrup

1. Gently heat the chocolate, butter, and corn syrup in a double boiler, mixing until it is well combined.

TO ASSEMBLE
1. Poke a hole in one end of each éclair shell.
2. the pastry cream to a piping bag with a tip large enough for the cream to move through.
3. Holding the éclair shell in one hand, place the tip of the piping bag to the hole and fill the shells with cream. You will feel the weight of each shell get heavy.
4. Gently dip the top of the filled shells in the chocolate glaze. In a rocking motion, dip from end to end, sealing the hole with the chocolate. If the shells are a little flat, cut them in half, add the pastry cream to the bottom half, and dip the top half outside in the chocolate and place the top half on top of the pastry cream.

Rule of thumb: Adding other ingredients to eggs interferes with the shrinking action of protein at higher temperatures, making it possible to boil a custard that contains even small amounts of flour or cornstarch. Same for lemon curd (p. 141).

LEMONY LEMON MERINGUE PIE

MAKES ONE PIE

There are three components to making a lemon meringue pie: crust, lemon curd, and meringue. The best lemon meringue pie I ever had was in Bridgewater, Nova Scotia, at the Light Delight Café. The crust was flaky and light, the curd was tangy, and the meringue was triple what you might expect from a boxed retail product. That pie is just a fleeting memory from the late '80s. As I fondly recall, it was the perfect reward at the end of the work week.

PASTRY

Buy a premade pie shell at the grocery store, thaw it out, and put it into your own pie dish or keep it in the foil pan. Or use the following recipe, which is my mother-in-law, Lucy's, signature pastry recipe. It makes two shells. You need only one pie shell for the bottom, so you can store the other in the freezer.

2 ½ cup flour
2 tsp baking powder
1 egg, beaten with 3 tbsp water
2 tsp sugar
¾ cup soft margarine or softened butter

1. Mix together the dry ingredients. Cut in the butter, add the egg mixture, and knead the dough gently until it is a smooth ball.
2. Divide the dough in two and wrap each half in plastic. Flatten the dough and allow it to rest in the fridge for 30 minutes.
3. Roll out one portion of the dough on a floured surface.
4. Fold the pastry in half with the fold in the centre, and transfer it to the pie plate.
5. Unfold the dough, fitting it into an ungreased pie plate, and press the dough out gently toward the centre.
6. Crimp the edges of the pie crust. Refrigerate for ½ hour.
7. Preheat the oven to 450°F.
8. Prick the pie dough all over with a fork.
9. Bake for 8–10 minutes or until golden brown. Cool.

LEMON CURD

4 egg yolks
1 ¾ cup sugar
¼ cup cornstarch
3 tbsp flour
2 cups water

1 tbsp lemon zest, finely grated
½ cup lemon juice
1 tbsp butter
¼ tsp salt

1. Separate egg yolks from 4 large eggs, reserving the egg whites for the meringue.
2. In a medium saucepan, mix the sugar, cornstarch, flour, and salt together.
3. Add the water and bring to a boil.
4. Whisk the egg yolks until slightly thick and light yellow.
5. Add ¼ cup of the hot liquid to the egg yolks, and then stir this mixture back into the saucepan.
6. Return to the heat and cook on low heat for about 5 minutes, stirring occasionally.
7. Remove from heat, and stir in the lemon juice, lemon zest, and butter.
8. Pour the lemon curd into the cooked pie shell.

MERINGUE

4 egg whites
½ cup sugar
¼ tsp cream of tartar

1. In a medium bowl, beat the egg whites with the cream of tartar on medium speed until frothy.
2. Gradually add the sugar, 2 tablespoons at a time, beating the mixture after each addition. At high speed, beat the mixture until stiff peaks are formed.
3. Spread the meringue over the lemon filling, carefully sealing in the edges of the crust and swirling the top decoratively. Bake for 7–9 minutes, until the meringue is golden brown.
4. Let the pie cool completely.
5. Cut with a wet knife.

CHOCOLATEY CHOCOLATE MOUSSE

MAKES 6 SERVINGS

This recipe is simple and easy and uses all the ingredients you usually have on hand. The coffee enhances the richness of the chocolate.

1 cup semisweet chocolate chips
1 tsp vanilla
4 eggs, separated

¼ cup strong coffee
1 tbsp butter
3 tbsp sugar

1. Combine the chocolate, butter, vanilla, and coffee in a heatproof bowl set over a pan of barely simmering water.
2. Heat until the chocolate is almost completely melted, stirring occasionally (or heat the chocolate mixture in a microwave for 1 minute).
3. Remove the bowl from the heat and stir until the mixture is smooth.
4. Whisk the egg yolks "to the ribbon" and add to the hot chocolate mixture, stirring vigorously until the mixture is slightly thickened.. Set aside and let cool slightly.
5. Whip the egg whites with sugar on medium-high speed until stiff peaks appear and a light meringue is formed.
6. Add one third of the whipped egg whites to the warm chocolate-yolk mixture, folding into the mixture to lighten it up.
7. Gently fold in the remaining egg whites with a spatula just until no visible streaks of egg white are left.
8. Divide the mousse into individual serving bowls and top with whipped cream.

FLAWLESS CRÈME BRÛLÉE

MAKES 4 SERVINGS

6 large egg yolks
1 ⅔ cup 35% whipping cream
¼ cup sugar
½ tsp vanilla

1. Preheat the oven to 325°F.
2. Vigorously whisk the egg yolks until they change colour and have thickened.
3. Add the sugar and vanilla. Whisk gently until the sugar is dissolved.
4. Add the cream and stir together.
5. Pour the mixture into ramekins, ½ cup per each.
6. Place the ramekins in a baking dish. Create a water bath by pouring boiling water around the ramekins until the water is halfway up the sides of the ramekins.
7. Place the baking dish in the oven and bake for 20 minutes until the custard is set. It should jiggle lightly.

Rule of thumb: In custards such as crème brûlée, where eggs are mixed with cream, the same principle applies: cooked at the right temperature, if the eggs are whisked vigorously, they thicken and add aeration, lightness and creaminess. If the eggs are whisked gently or stirred without aerating, they thicken and add firmness.

GRANOLA

MAKES 5–6 CUPS

This granola uses egg whites to add some elasticity and hold the granola together in clusters.

3 cups oatmeal (quick oats, minute, or other rolled oats)
1 cup unsweetened shredded coconut
1 cup walnuts, coarsely chopped
2 tbsp olive oil or lightly flavoured oil
½ tsp salt
⅔ cup maple syrup or maple-flavoured table syrup
½ tsp cinnamon
1 egg white

1. Preheat the oven to 300°F. Line a baking sheet with parchment paper.
2. Combine all the ingredients except the egg white and salt in a large bowl, tossing to coat everything evenly.
3. Whisk the egg white in a separate bowl, add the salt, and whisk until frothy.
4. Stir the egg white into the granola mixture, distributing it throughout.
5. Spread the granola in a single layer on parchment-lined baking sheet.
6. Bake for 35–45 minutes. About halfway through the baking time, use a large spatula to turn over sections of the granola carefully, breaking them as little as possible.
7. When the granola is evenly browned and feels dry to the touch, transfer to cooling racks. Cool completely.
8. Once the granola is cooled, break the granola into clusters.

Rule of thumb: You can substitute almonds for walnuts. If you want sweeter granola, increase the syrup to ⅔ cup. This granola keeps for 2–4 weeks at room temperature in a sealed container.

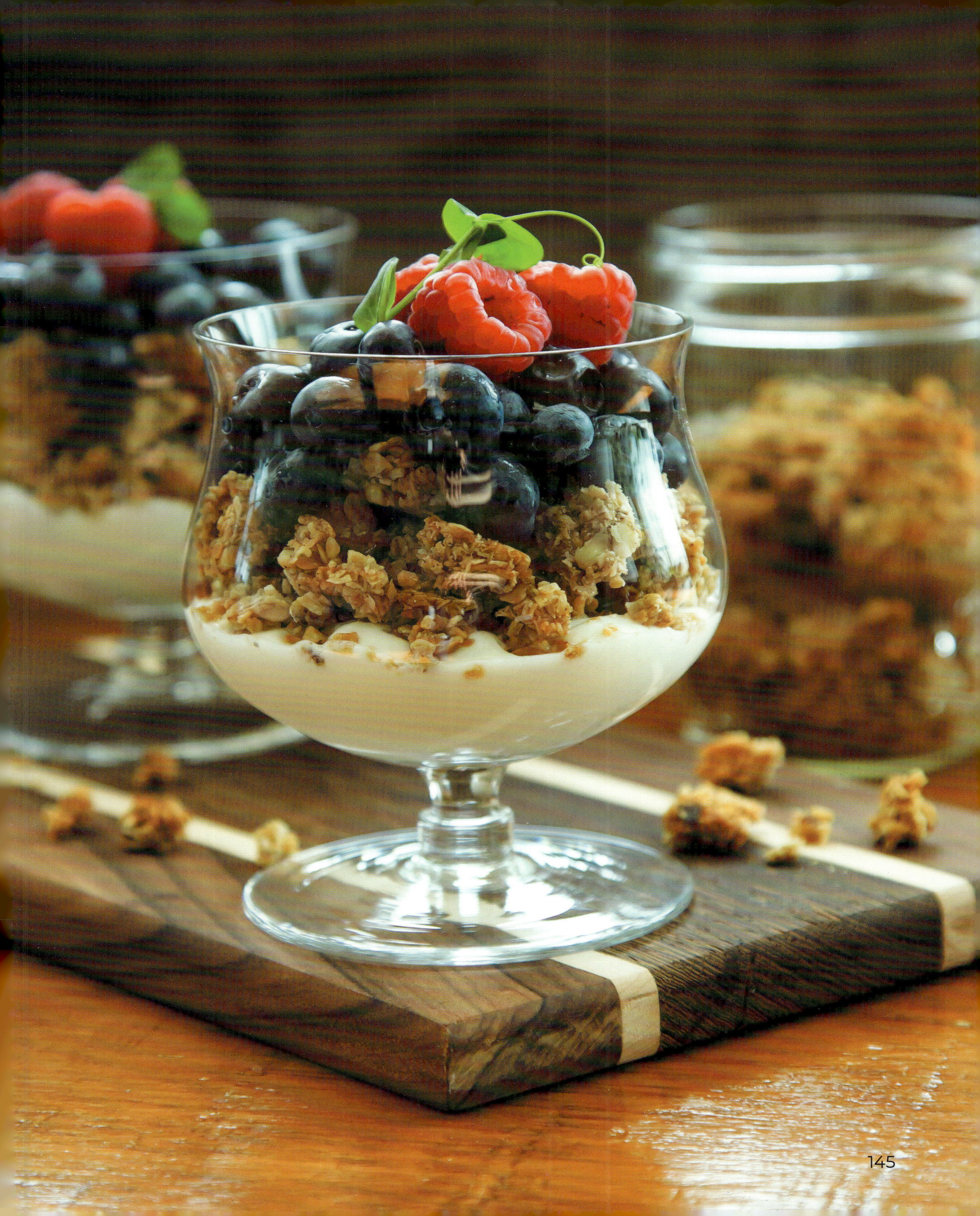

PAVLOVA

MAKES 12 SERVINGS

Pavlova is a glorious dessert prepared with a meringue base, topped with lemon curd, fruit compote and whipped cream. In one fork full, you experience delicate brittleness and sweetness from the meringue, smoothness and tartness from the curd, creaminess from the whipped cream and freshness from the compote. It is worth the effort.

A pavlova meringue is traditionally made by slow cooking aerated egg whites and sugar. Before cooking it, the meringue has a unique squidgy texture thanks to the addition of an acid such as cream of tartar (tartaric acid) and a little cornstarch. The same holds true for a vegan alternative. This is an interesting way of using the juice of canned chickpeas. The juice is called aquafaba, and you can aerate it like egg whites. One 540ml (19oz) can of chickpeas will give enough aquafaba for this recipe.

FOR THE MERINGUE USING AQUAFABA

¾ cups (190 g) aquafaba
¾ cup (165 g) sugar
2 tsp (10 g) cornstarch
¾ tsp vanilla
¼ tsp cream of tartar
⅛ tsp salt

1. Dissolve the sugar in the aquafaba (warm in the microwave if needed to speed up the process)
2. Add the cornstarch, vanilla, cream of tartar, and salt. Mix until combined.
3. Whip on high until the meringue is light coloured, thick and gooey and stiff peaks form, 9–12 minutes.

FOR THE MERINGUE USING EGG WHITES

1 ¼ cup (180 g) egg whites from 6 large eggs
1 ¼ cups (360 g) sugar
2 tbsp cornstarch
1 tsp vanilla extract
¼ tsp cream of tartar
⅛ tsp salt
1 tbsp lemon juice

1. In a large mixing bowl, mix the egg whites and sugar on low speed until the sugar is dissolved.
2. Add the cornstarch, vanilla, cream of tartar, lemon juice and salt. Mix until combined.
3. Whip on high until the meringue is white in colour, thick and gooey and stiff peaks form, 3-4 minutes.

COOKING INSTRUCTION FOR MERINGUES

1. Preheat the oven to 225°F and line two large baking sheets with parchment paper.
2. Using a large tip, pipe a swirl with four circles starting from the inside out. Double up the last round to create a ridge that will hold in the curd, cream and compote (or berries).
3. Bake for 1 hour 15 minutes and then turn off the oven. Let cool for an hour in the oven.
4. Remove the meringues from the oven. Cool them and store them in an airtight container.
5. Store the meringues in the fridge for up to two weeks or in the freezer for three months.

Rule of thumb: Add a little salt to the egg whites or aquafaba to activate the proteins and accelerate the aeration. The aeration of the aquafaba takes twice as long as egg whites.

Salt also enhances the overall flavour while tempering the sweetness of a lot of sugar.

Meringues are dried so they are crisp but they will pull humidity from the air and become sticky on the outside and soft if left out. Separate each meringue circle with a piece of parchment paper and store in an airtight container in the fridge for up to two weeks.

TO ASSEMBLE

2 cups (500 g) 35% whipping cream
1 cup (125 g) icing sugar, plus extra to dust over the finished pavlovas
2 tsp vanilla extract

12 meringues
1 ½ cups lemon curd (see p. 141)
1 ½ cups berry compote (see p. 228)
fresh berries and mint leaves for garnish

1. In a large cold mixing bowl, add the whipping cream, icing sugar and vanilla. Whip on high until thick and stiff peaks form.
2. Place a meringue in the center of a nice dessert plate or shallow bowl.
3. Top with ¼ cup whipped cream, 2 tbsp lemon curd, and 2 tbsp compote.
4. Dust with icing sugar (optional) and garnish with mint and berries.

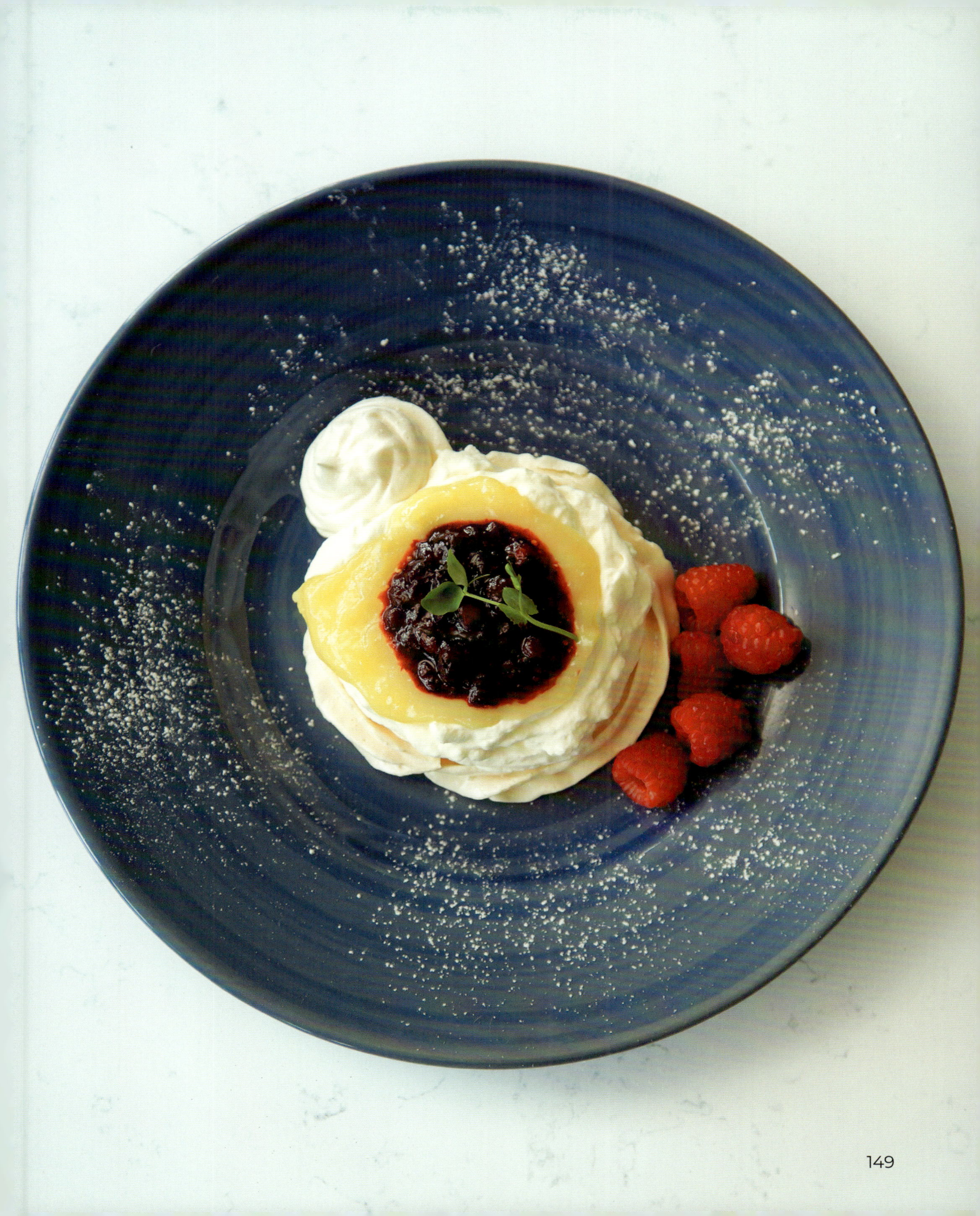

150

8. TA DA! Fish and Seafood

MY STORY: SEEING IS BELIEVING

My parents have a cottage in the small rural community where my father grew up. It is, of course, on the water, so I grew up around fishing, boats, and fresh seafood. Cod has always been the staple fish for every Newfoundlander, and we were no different. Catching cod was not only a favourite pastime but also a means to supper and, for my father and me, careers in the seafood industry. My first food science university course introduced me to what happens biochemically when you catch a fish and end up having it for supper. It was a pinnacle course. My father and I went cod jigging shortly after I read a white paper on post-mortem physiology. I was wide-eyed when I observed the process happen in front of my eyes that day. We caught the fish, cut it, and gutted it and watched it stiffen into rigor mortis and relax out of it over the course of that day. We then filleted and skinned the fish, and pan-fried the fillets, savouring the delicate flavour and perfect white flakes. What impressed me was knowing the science behind what happened and how it ultimately assures the absolute best-quality food. That experience and the white paper on the biochemical process that converts fish into food became my passion for over 30 years as a meat, poultry, and seafood scientist who created thousands of food products for Canadians. I still have a copy of that paper.

THE MAGIC OF SCIENCE: FISH IS NOT SMELLY

It's a common misconception that all fish and seafood smell bad and are slimy, so people struggle to add them to their diet. However, good-quality fish and seafood do not smell fishy and are not slimy. You can expect a slight sweet odor and a little moistness to the flesh, if it is fresh and wholesome. The secret to keeping fish and other seafood fresh comes down to processing and how it is stored. Like most food, once fish is harvested, it begins to deteriorate. There are a few techniques to maintain its quality and wholesomeness and to slow down the degradation. Fish and seafood are most susceptible to deterioration and can be highly perishable if not handled well.

When fish try to escape a hook or net, they expend a lot of energy. As they do so, their muscles accumulate lactic acid, which is the sweet aroma you smell. To reduce acid accumulation and deterioration of the muscle, the fish should be bled quickly. Doing this ensures the muscle can hold moisture and maintain its fleshy structure. Although the muscle naturally stiffens (i.e., goes into a state of rigor mortis), it then relaxes, and natural enzymes take over and age the meat. Holding fish in a refrigerator or freezer slows the aging process, maintaining this wholesomeness.

The least offensively fishy fish to start off with is mild cod, haddock, or halibut. Some argue tilapia is the mildest fish, and it does have a mild sweetness. Tilapia can be substituted in many of the recipes in my book, even though it is a little softer than cod, haddock, or halibut. Interestingly, these fish can be substituted in most chicken recipes too. Try it!

Arctic char, rainbow trout, and Atlantic salmon are most widely available but have their own distinct flavours and textures, because of their high omega 3, 6, and 9 content. These fish are worth trying but check for the sweet smell and no slime.

Shrimp has very wide acceptance and is quick and easy to prepare. It can absorb other flavours and complements dishes with its own distinct flavour.

COOKING FISH AND SEAFOOD

There really are only a few ways to cook fish and seafood, and they generally don't take much time. All fish and all seafood require just a few minutes of cooking. The delicate nature of fish and seafood makes it a speedy meal for even the busiest of people. Most fish are pan-seared and then finished in the pan (pan-fried) or the oven (roasted) to heat the flesh all the way through. A ton of fish-and-chip shops around the world deep-fry fish because it is fast and flavourful. Poaching and steaming use hot liquid to cook the flesh of fish and seafood respectively.

Pan-searing is an incomplete process, a first step in a larger process. Searing is a surface treatment used to quickly create a flavourful brown crust on protein cuts. Searing is often required before roasting, braising, or finishing cuts of meat as well. Here are the steps involved in pan-searing:

1. Season your protein well on both sides with salt and pepper.
2. Place a cast-iron frying pan or heavy non-stick pan on the burner of your cooktop.
3. Turn the heat to high and add 2–3 tbsp of oil (enough oil to moisten the protein but not have it swimming in the fat). You can also use a combination of oil and butter as an alternative to oil.
4. Once the oil is lightly smoking or a drop of water spits and sizzles when added to the pan, add your protein.
5. Immediately reduce the heat to medium to ensure the protein does not burn. Flip it over as soon as you see a golden-brown crust forming and brown the other side too.
6. Finish cooking either on the stove or in the oven at 400°F for a few minutes until the protein is opaque and reaches the desired cook temperature (see p. 7).

Pan-frying is an extension of the pan-searing process where the fish is seared and then the heat is reduced, and cooking is finished in the pan. It is ready to serve. As an aside, some seafood, like shrimp, are often cooked by sautéing, where the shrimp is moistened with oil or fat and, on medium-high heat, "jumps" in the pan. Butter is usually used for sautéing to flavour the food, but nowadays we use light-bodied oils like sunflower or corn oil, which can be heated to high temperatures.

Poaching and steaming are two cooking methods that make use of hot moisture to cook protein. With poaching, either submerge directly in water or use a seasoned broth and allow to simmer until the flesh is opaque. To steam the fish, surround it with steam from other vegetables or a pan of water beneath the fish, allowing the heat from the steam to transfer to the flesh of the seafood.

- Glazed Pan-Seared Salmon
- Riesling-Poached Halibut
- Seared Scallops with Garlic Lemon Cream Sauce
- Pan-Fried Cod with Scrunchions
- Fish Cakes with Million-Dollar Relish
- Beer-Battered Cod
- Baked Stuffed Squid
- Lobster Linguine Tutto Mare
- Mussels

GLAZED PAN-SEARED SALMON

MAKES 4 SERVINGS

4 6-oz salmon fillet portions with or without skin, fresh or thawed
Salt and ground black pepper

2 tbsp light tasting vegetable oil
A knob of butter (approximately 2 tbsp)
½ cup chili peach glaze

1. If your fish has skin, scrape the scales off and wash well.
2. Cut 4–5 slits in the skin of a portion of salmon, and give both sides of the salmon a generous dusting of salt and pepper.
3. Place the salmon in a very hot pan with a little oil, searing the fish until a brown crust develops on the underside.
4. Flip and sear until the other side is browned.
5. Finish by adding a knob of butter to the pan and basting the salmon for a minute or two to cook it all the way through in the pan. Or finish the salmon in a preheated oven at 400°F for 3–5 minutes until the fish is done all the way through, firm to the touch, and reaching an internal temperature of 158°F (71°C)
6. Serve the salmon immediately, and drizzle with chili peach glaze.

CHILI PEACH GLAZE

MAKES ½ CUP

1 ½ tbsp honey
¼ cup peach cantaloupe preserves
2 tsp grainy Dijon mustard
2 tsp soy sauce

2 tsp lime juice
½ tsp minced garlic
¼ tsp chili flakes

1. Mix all the ingredient together in a small saucepan.
2. Bringing the mixture to a boil, then simmer the glaze until it thickens slightly and coats the back of a spoon.

RIESLING-POACHED HALIBUT

MAKES 6 SERVINGS

This recipe is prepared with halibut, a firm whitefish. The sweetness of the Riesling and the lemon enhance the delicateness of the fish. I have used this poaching liquid and cooking method with all types of white-fleshed fish. It is very versatile, and the directness of the cooking method heats the fish quickly and uniformly.

POACHING LIQUID

1 cup water
1 cup Riesling
¾ cup diced onions
¾ cup chopped celery
1 lemon, sliced

1. Bring the water and wine to a simmer in a saucepan with the onion, celery, and lemon.
2. Simmer for 15 minutes, strain, and keep warm.

FISH

1 tbsp butter
½ onion, finely chopped
6 pieces halibut fillet
Salt
1 tbsp fresh tarragon (or 1 tsp dried)
2 tbsp chives, thinly sliced
1 lemon, cut into 6 wedges

1. Melt the butter in a large, shallow pan over medium heat.
2. Sauté the onions until translucent, about 5 minutes.
3. Season the fish with salt and add to the pan.
4. Add poaching liquid and tarragon, then simmer gently for 8 minutes or until the fish is firm to the touch and white in the centre.
5. Remove the fish carefully with a slotted spoon.
6. Top with chives and a little poaching liquid. Serve with lemon slices.

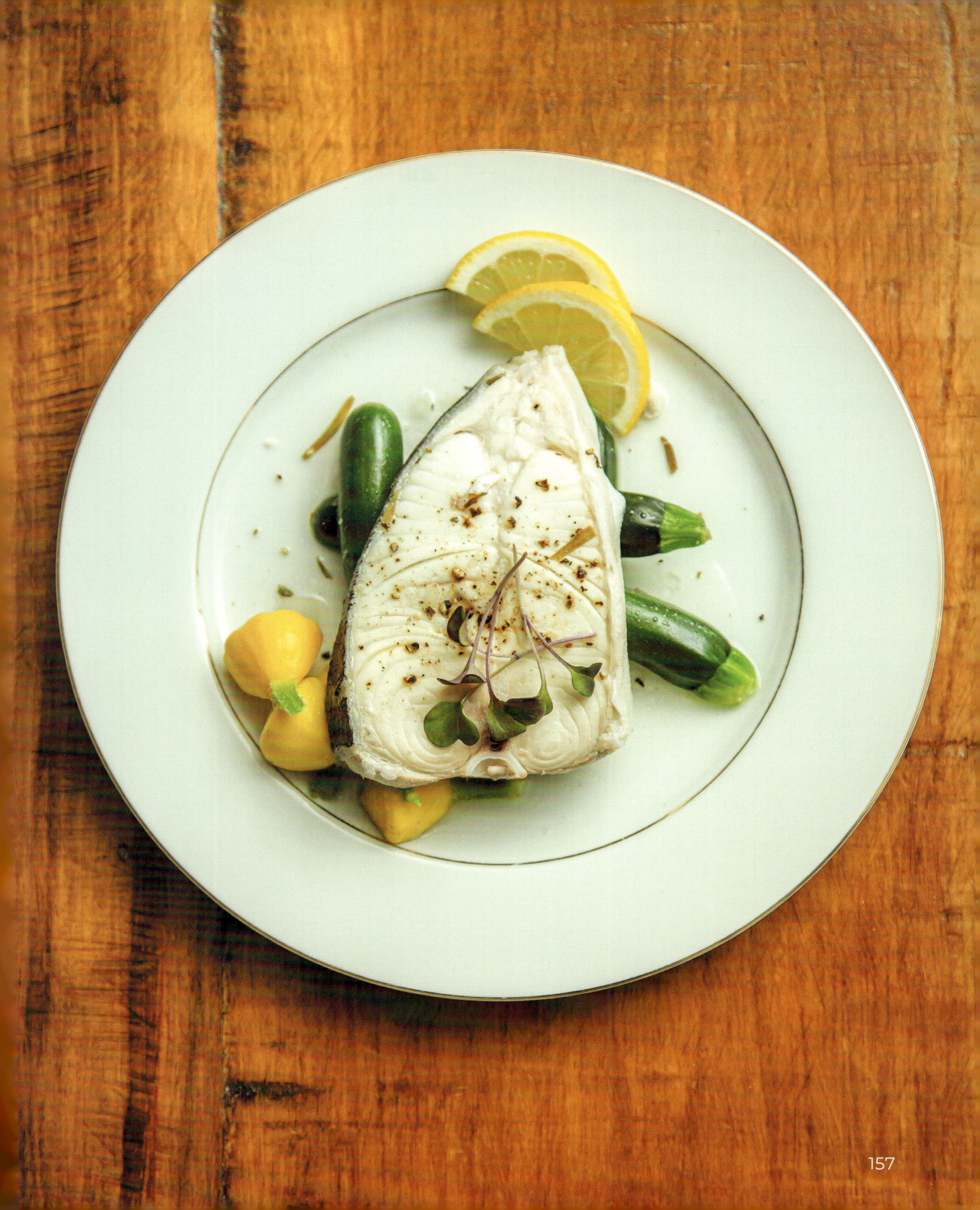

SEARED SCALLOPS WITH GARLIC LEMON CREAM SAUCE

MAKES 3-4 SERVINGS

12 sea scallops (10 count size)
2 tbsp light-tasting vegetable oil
Coarse salt and ground pepper

Scallops are sold based on the number or count per every 1lb (454g). The lower the count, the larger the scallops.

SAUCE
¼ finely chopped shallots or onions
1 tbsp lemon juice
½ tbsp lemon zest (from ½ lemon
⅔ cup light-bodied white wine
or vegetable broth

1 cup whipping cream
1 tbsp chopped chives
2 tbsp finely diced red peppers

1. Remove the scallops from the fridge and pat them dry.
2. Heat a large non-stick frying pan over high heat. Add the oil.
3. Season both sides of the scallops with salt and pepper.
4. Place the scallops in the pan only when the oil is swirling and a drop of water spits and sizzles when dropped in.
5. Jiggle the scallops slightly with a pair of tongs to loosen them from the pan, and cook 3 minutes on each side.
6. Remove the scallops and set aside.
7. Make the sauce by adding in the garlic, wine (or broth), and lemon juice to the frying pan. Let the mixture bubble for a minute.
8. Add the cream, and simmer the sauce until it reduces so the sauce sticks to the back of a spoon.
9. Add in the scallops and let them just heat through. The middle of the scallops should be somewhat translucent when you cut into them.

Rule of thumb: When searing scallops or any fish or seafood, it is better to sear a few at a time, leaving 1" space between them.

Overcrowding the pan just cools it down, and the scallops don't have the chance to seal in the juices. The juices end up being released, and the scallops become small, tough, and rubbery.

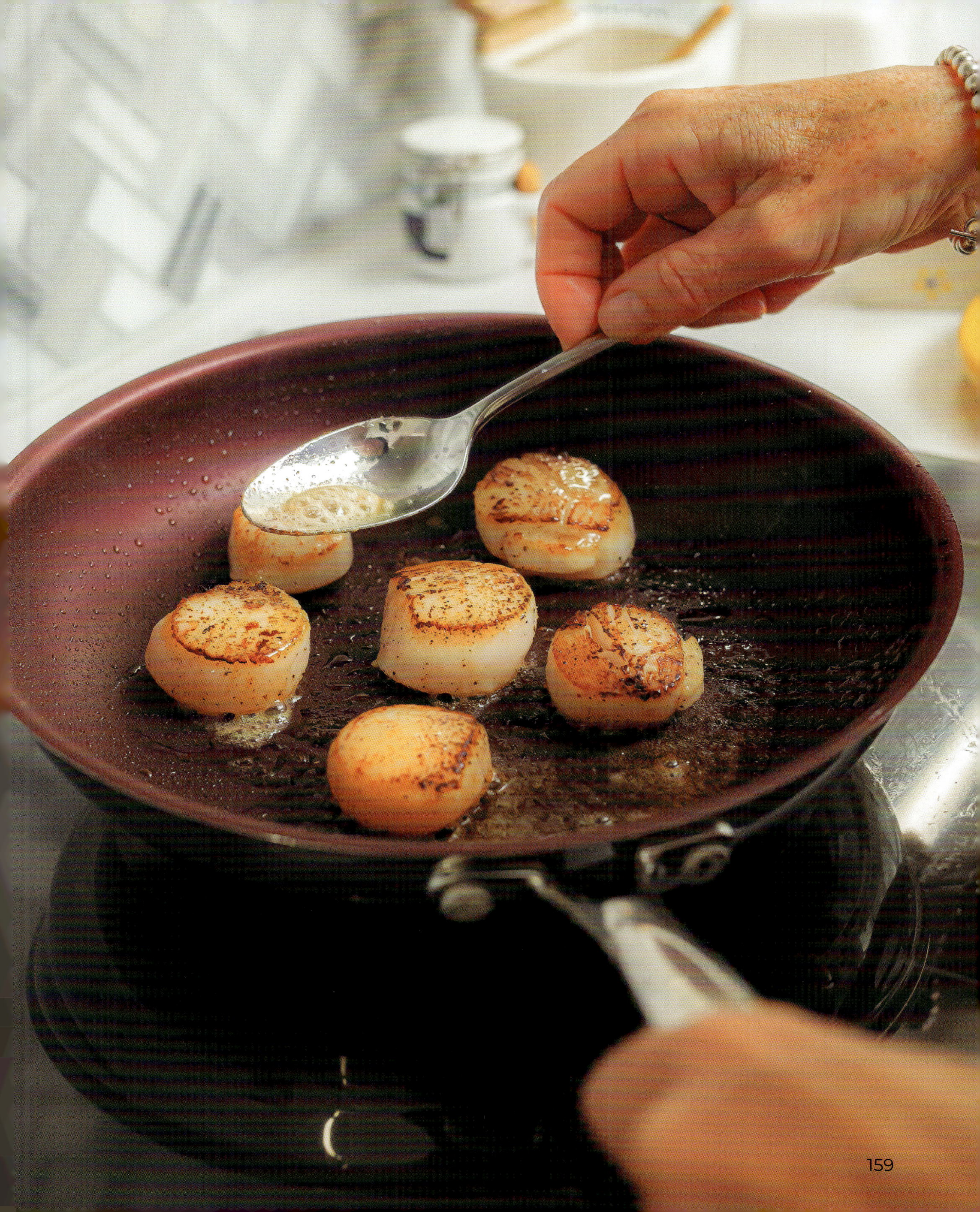

PAN-FRIED COD WITH SCRUNCHIONS

MAKES 4–6 SERVINGS

The heat and salt from salted pork backfat imparts flavour and the energy needed to quickly heat through the dense flesh of the cod fish. This fast heating is important when large families need to be fed quickly. The fat lasts a long time when packed in salt, particularly in rural areas that are isolated during winter months. Scrunchions are the result of rendering small, diced pork fat to a crispy nugget.

4 cod fillets
2 tbsp flour
Salt and pepper

Salted pork backfat, rind removed and cut into ¼" cubes,

1. Dredge the fillets in flour that has been seasoned with salt and pepper.
2. Heat a cast-iron pan on high. Add the pork fat and cook until there are crispy pieces of fat and a pool of fat.
3. Add the fish to the fat. Turn the heat to medium and fry the cod until the flesh is flaky.
4. Remove and serve with the crispy scrunchions, ladling the fat over top of the fish.

FISH CAKES

MAKES 4-6 SERVINGS (ABOUT 12-16 FISH CAKES)

Every Newfoundlander has their own fish cake recipe and believes it to be the best. If you are from the east side of the island, you likely add savoury, a special summer savoury from Mount Scio in St. John's. This recipe is the one we prepare at Java Jack's Restaurant & Gallery, and yes, we use savory. It's perfect paired with our Java Jack's Million-Dollar Relish. Order it online (see p. 256) or make your own.

1 medium onion, diced ¼"
¼ cup butter, salted or unsalted
1 lb frozen cod fish, tails, loins, or trims
½ cup frozen salt cod pieces, boneless and skinless

5-6 medium boiled potatoes
1 tbsp Mount Scio savoury
2 tbsp all-purpose flour or cornstarch
¼ cup light tasting vegetable oil
Salt and fine black pepper to taste

1. Melt the butter in a frying pan or pot. Add the diced onions and cook the onions until they are translucent but before there is any browning. Set aside.
2. Cook all the fish pieces together in water. Strain them very well until there is no water dripping.
3. Cook the potatoes until soft and ready for mashing. Strain very well until there is no water dripping.
4. Mash the potatoes. Add the cooked fish and onion-butter mixture.
5. Mix all the ingredients long enough that there are no lumps, and it is homogeneous and smooth. The final mixture should have a cookie-dough consistency.
6. Add the savoury, salt, and pepper. Mix and adjust to taste.
7. Portion the fish-potato mixture into 56 g (2 oz) circular cakes.
8. Roll each cake in the flour or cornstarch. Dust off the excess.
9. Heat a frying pan on high. Add the oil 2 tbsp at a time.
10. Fry the fish cakes in the oil until the cakes are brown on each side. Add more oil so the cakes don't stick to the pan.
11. Serve the fish cakes hot with a dollop of our Million-Dollar Relish.

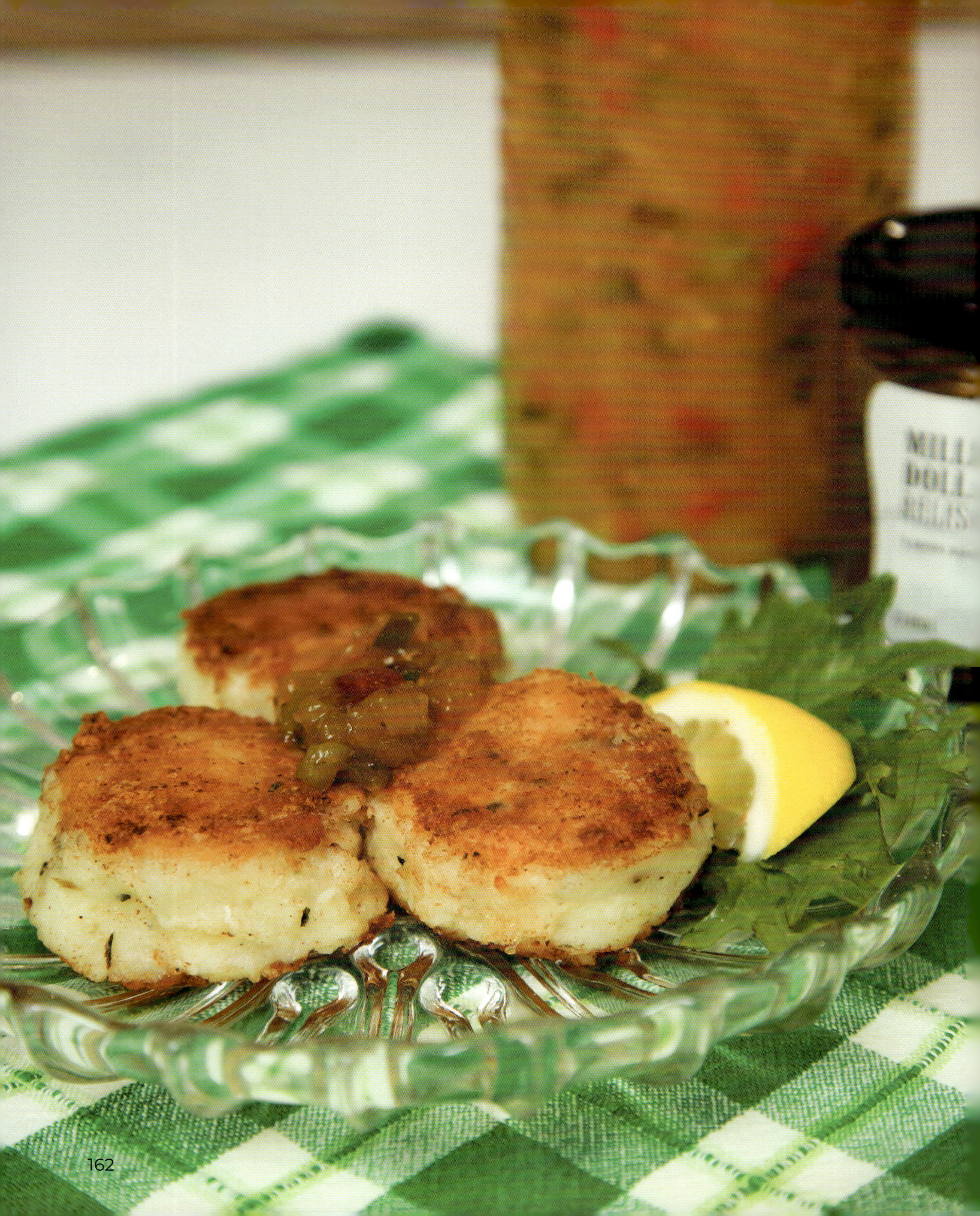

MILLION-DOLLAR RELISH

MAKES ABOUT 10 CUPS (2 L)

This is the Million-Dollar Relish recipe that has been an anchor ingredient at Java Jack's since the beginning. It is the perfect complement to any fish dish or burger.

3 large English cucumbers, diced ¼"
2 lbs onions, diced ¼"
1 ½ medium green peppers, seeded and diced ¼"
1 medium red pepper, seeded and diced ¼"
½ cup salt
8 cups cold water

2 cups white vinegar (5% acetic acid)
3 cups white sugar
½ tbsp turmeric powder
½ tbsp celery seeds
½ tbsp mustard seeds
2 tbsp cornstarch
2 tbsp water

1. Make a brine solution by adding the salt to the cold water. Stir until the salt is dissolved.
2. Add the chopped vegetables to the brine, and let them soak overnight completely submerged.
3. Drain the vegetables in the morning and rinse off the excess brine from the vegetables.
4. Place the vegetables in a large stock pot
5. Add the vinegar, sugar, turmeric, celery seeds and mustard seeds
6. Mix everything together until it's well combined.
7. Bring the mixture to a boil, then turn down the heat to simmer and cook for 1 hour.
8. Mix the cornstarch and water together until there are no more clumps. Add this mixture to the pot while stirring vigorously. Cook until the mixture is thick enough to adhere to the back of a spoon.
9. Cool the mixture slightly, but only enough so not to burn yourself. Fill cleaned, sterilized glass jars with hot relish, leaving just ½" headspace. Cover tightly with the lids. Let the relish cool until the lids create a vacuum and give a little "pop" sound. Store in a cool location for up to six months. It lasts for six months in the fridge after opening.

Rule of thumb: *The acidity in this relish and the percent of solids such as the sugar keep the pH and available water low enough to preserve the vegetables and inhibit growth of harmful bacteria that spoil foods and can make us sick.*

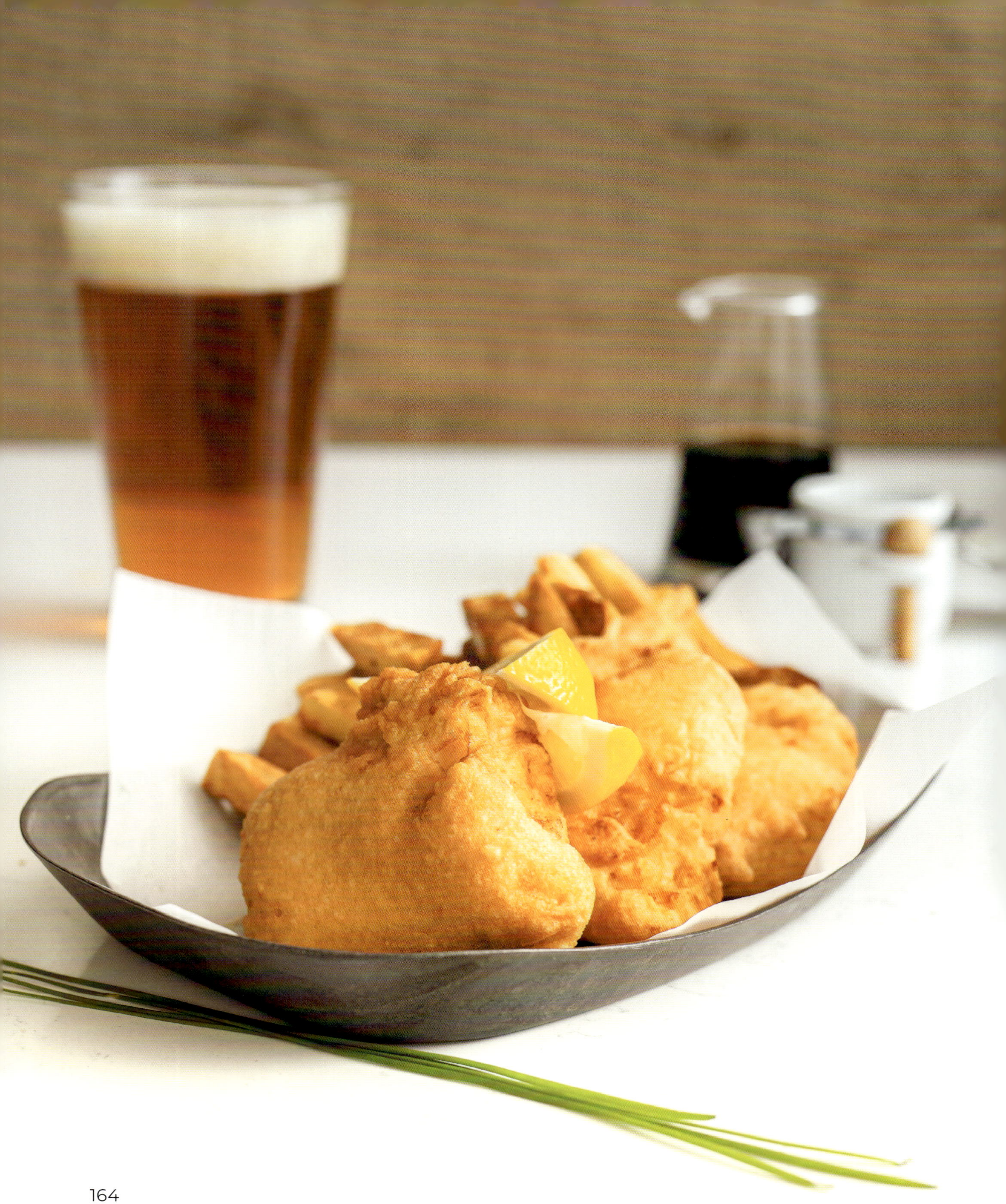

BEER-BATTERED COD

MAKES 4 SERVINGS

Deep-fried anything is not my style, but once a year we dip a piece of fish, usually cod, in this delicious, tender Orly beer batter—a batter leavened with beer and fluffy whipped egg whites. The batter is set by deep-frying. We douse the finished cod pieces with malt vinegar and a little sea salt and serve with my own coleslaw (see p. 183). It's very good!

ORLY BEER BATTER

1 cup (150 g) flour
2 egg yolks
⅔ cup (approximately ½ bottle) light ale or lager (an IPA is too bitter)
2 tbsp light tasting vegetable oil
½ tsp salt
2 egg whites, stiffly beaten

1. Place the beer, oil, and egg yolks in a bowl. Whisk together well.
2. In a separate bowl, mix the salt and flour.
3. Gradually incorporate the flour mixture into the beer, oil and egg yolk mixture to ensure a smooth consistency.
4. Beat the egg whites until stiff and fold them into the batter mix. Try not to overwork the batter as it makes it leathery.

FISH

4 6-oz pieces cod loins, fresh or thawed and cut in half
Flour for dusting (about 2 tbsp)
8–12 cups light tasting vegetable oil for frying

1. Dry each piece of fish with a towel or paper towel to remove the excess water.
2. Dust each piece with a little flour.
3. Preheat the vegetable oil in a safe certified deep-fryer to 375°F.
4. Dip each piece of fish in the batter and let the excess batter drip off. Gently hold the battered cod over the hot oil and dip it slowly into the oil.
5. Cook each piece for about 10 minutes, until the exterior is light golden brown and the fish turns from translucent to opaque inside.

Rule of thumb: Hold the battered piece over the oil, dipping a corner or small portion so it can cook and seal before letting the whole piece submerge in the oil. Doing this prevents the fish from sticking to the wire mesh basket in the oil.

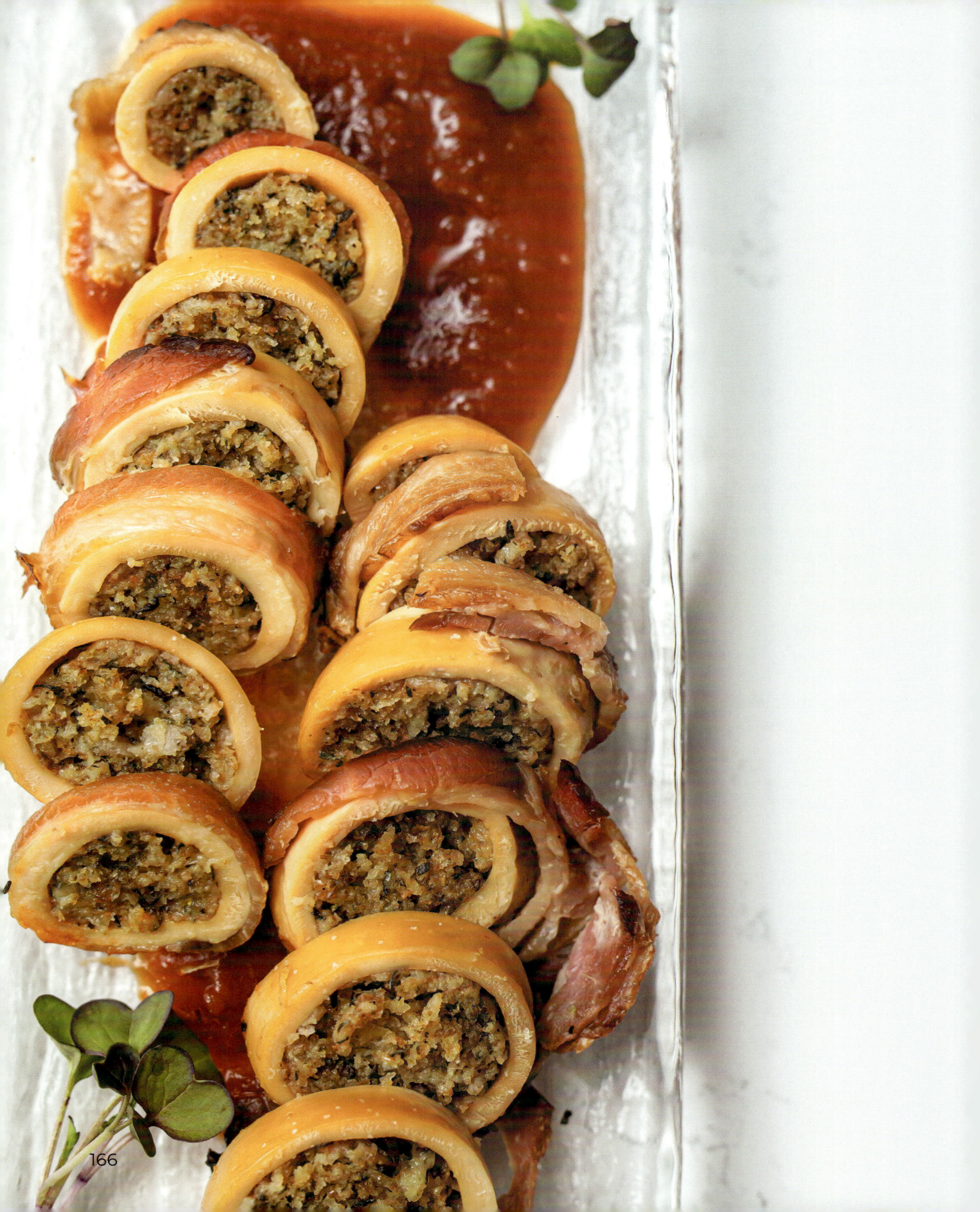

BAKED STUFFED SQUID

MAKES 4–6 SERVINGS

Squid, also known as calamari by many these days, was considered a by-product but has gained popularity. Deep-fried calamari rings are a common appetizer in pubs and restaurants. Here you can make it without deep frying and consider serving it as part of a charcuterie board.

6 squid tubes, cleaned
3 cups traditional bread stuffing
6 bacon slices
½ cup broth or tomato juice
½ tbsp flour

Rule of thumb: The secret to tenderizing squid tubes is to use humidity in the cooking. Hot temperatures and high humidity combine to break down cross links in the squid's collagen.

1. Preheat the oven to 375°F.
2. Wash the tubes.
3. Mix all the stuffing ingredients together.
4. Grease a 9"x11" baking dish with cooking spray oil. Pour the vegetable broth or tomato juice in the bottom, enough to create a puddle for the squid to sit in.
5. Fill the tubes with bread stuffing, wrap in a slice of bacon, and place the filled, wrapped tubes in the baking dish in the puddle of liquid.
6. Cover the baking dish completely with foil and bake, covered, in the oven for 1 hour. Remove the cover and check the tenderness of the squid – a fork should push easily through the squid tube. If all the liquid is absorbed, add some additional warmed broth or juice.
7. Continue to bake the squid, covered, until they are tender. Remove the cover and bake for an additional 15–20 minutes to crisp the bacon.
8. Remove the stuffed squid and strain the tomato juice.
9. Add the flour and whisk on high heat until it's thickened.
10. Pour this thickened juice onto a serving dish.
11. Slice the baked squid and serve over tomato sauce.

HOW TO MAKE TRADITIONAL BREAD STUFFING

Let your leftover buns and loaf ends dry out to make bread crumbs. You'll need 3 cups of bread crumbs in total. When your bread is dried out, grind it into crumbs. Add the three cups of bread crumbs to ¼ cup melted butter, 2 tbsp minced onion, ½ tsp salt, ¼ tsp black pepper and 2 tsp Mount Scio savoury (a summer savoury). Combine everything until all the crumbs are dampened and the seasoning is distributed evenly throughout the mixture. If you're making the stuffed squid recipe, use this bread stuffing mixture in this uncooked state. (See also p.??)

LOBSTER LINGUINE TUTTO MARE

MAKES 2 SERVINGS

I discovered this Java Jack's signature dish while practising for my practical chef exam. *Tutto mare* means "everything from the sea," so it seemed appropriate to use the essence of this recipe to showcase the fresh sweetness of lobster straight from local fishermen. The flavours of the vegetables, light-tasting olive oil, dry white wine, and homemade lobster stock cling nicely to al dente linguine and meld together so well with the steamed lobster meat from the claws. The dish is presented with a butterflied lobster tail on top like a gift. It's a special dish you can make for yourself at home. Give it a try!

100 g linguine
Enough salt to salt your pasta water
¼ cup dry white wine
½ cup carrots, diced ¼"
½ cup onion, diced ¼"
3 tbsp extra-virgin olive oil
½ cup cherry tomatoes, halved
½ cup mushrooms, sliced
2 tsp garlic, chopped or minced
3 tbsp parsley, finely chopped

¼ cup vegetable stock or shellfish stock, preferably lobster stock made from lobster bodies
1 ½ cups lobster meat, cooked and cut into bite-size pieces
2 lobster tails, uncooked, fresh or thawed
2 nobs of butter (about 4 tbsps)
2 tbsp Parmigiano Reggiano shaved for garnish

1. Boil a pot of water. Add a little salt. Add the dry linguine and cook until it is al dente. (Follow the time provided on the package directions.) Drain the pasta and toss it with 1 tbsp of the olive oil, and set aside.
2. Heat the rest of the olive oil in a large frying pan and sauté the onion and carrot until soft.
3. Add the mushrooms and garlic
4. Lower the heat and cook for 2–3 minutes.
5. Add the white wine and reduce the volume on medium heat for 4–5 minutes.
6. Add the tomatoes. Simmer for 5–7 minutes.
7. Add the lobster meat and simmer for 3–4 minutes.
8. Add the linguine and toss everything together to heat it.
9. Taste and season with salt and pepper.
10. Plate the linguine, lobster, and vegetables, and garnish with a few shavings of Parmigiano Reggiano and a little parsley. Top with the cooked butterflied lobster tail (see above).

HOW TO PREPARE LOBSTER TAIL
Lay the tail on the counter. With a pair of scissors, cut the back centre of the tail down to the fins. Then turn the tail over. Skipping the first rib at the opening, snip all the other rib bones on the underside. Turn the tail right side up and pull the raw lobster meat from the inside with a fork or use your fingers to pull the meat out gently. Keep the small fin portion attached. Place the tail on a broiler pan and top each tail with a nob of butter (about 1 tablespoon). Cook at 400°F for 10 minutes or until the lobster meat is opaque but not hard or rubbery.

There is no love sincerer than the love of food.
-George Bernard Shaw

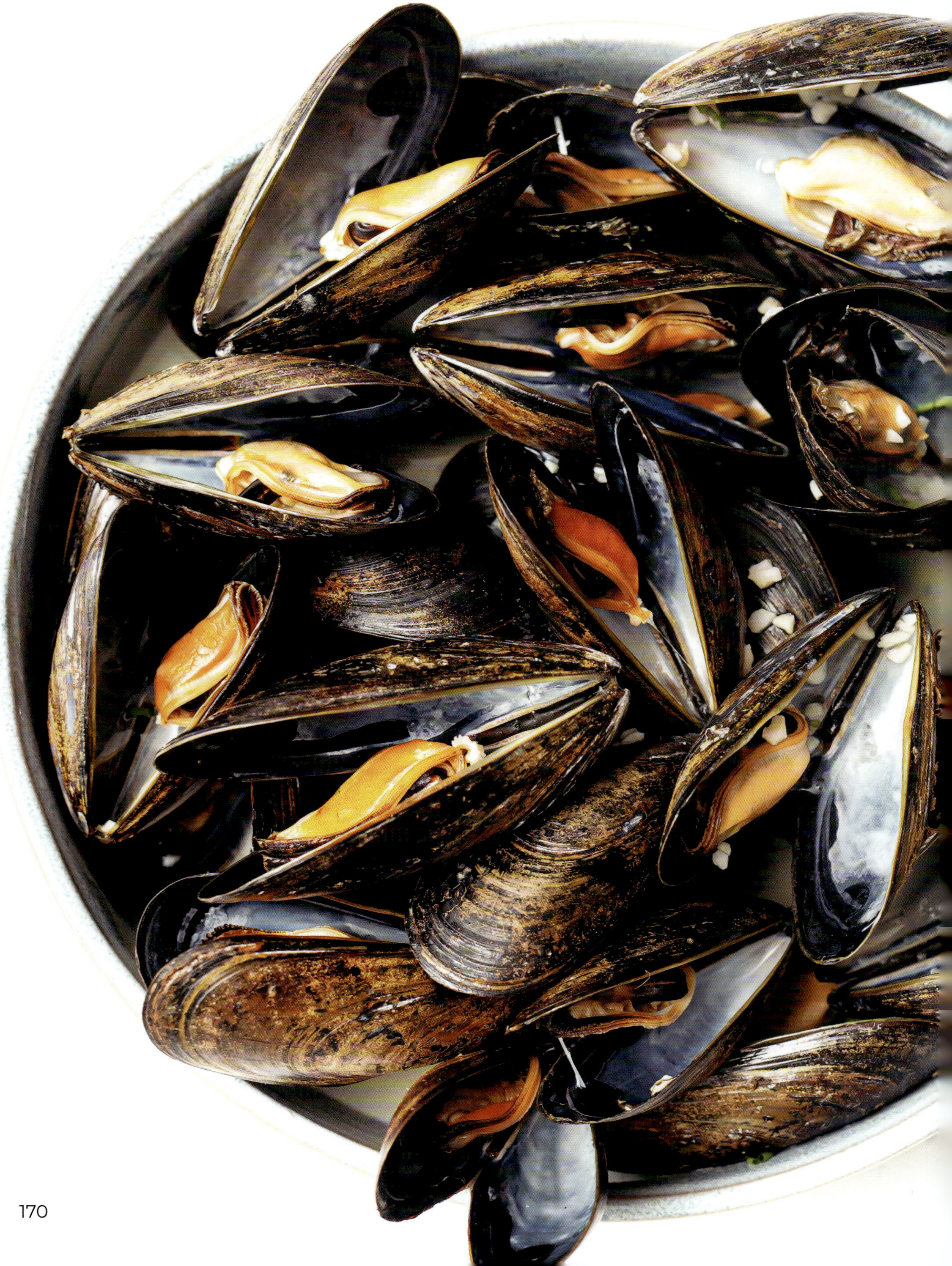

MUSSELS

THIS IS A GREAT APPETIZER FOR 4

All you need is:
5 lbs mussels
A liquid to cook the mussels in
Seasoning or herbs

1. Place the mussels and cooking liquid in a large pot or frying pan with a cover.
2. Ensure half of the pot is available without liquid for the mussels to open.
3. Steam for 5 minutes until all the mussel shells are open.
4. The mussels are flavoured and ready to eat.

The muscles of the mollusc shrink and the shells pop open with the steam.

CLASSIC MUSSELS
½ cup white wine
1 tsp chopped or minced garlic
Garnish with parsley

PESTO CREAM MUSSELS: A JAVA JACK'S FAVORITE
½ cup cream
1 tsp basil pesto
Garnish with basil and pine nuts

MARINARA MUSSELS
½ cup marinara sauce
Garnish with parsley and oregano

MARGARITA MUSSELS
½ cup tequila
Lime
Marinara sauce

9. Magic in Meat

I'd be remiss if I did not include some fundamentals of meat science in this book. Nations around the world consume skeletal muscle as meat as a main source of protein and energy. Other countries are adding more meat to their diets as their economies prosper. As a food scientist who specialized in meat science, with 30+ years of experience working in the meat industry, who better to share this information? Knowing what is in meat and its vast potential in functionality allows you to create the tastiest, most succulent, nutritious, centre-of-the-plate options in your cooking repertoire.

MEAT AND ITS COMPOSITION

"Meat" is the flesh of animals and fish used as food. It plays a very significant role in our diets. The principal merit of meat as a food is its high digestibility and nutrient content. It contains essential amino acids in the form of muscle protein; water-soluble vitamins such as thiamine, riboflavin, and niacin; some minerals such as copper, iron, and phosphorus; and high-energy fats and lipids, including essential fatty acids.

Meat is composed primarily of muscle, plus variable quantities of all types of connective tissue, as well as some epithelial and nerve tissues. Skeletal muscle is the main source of muscle tissue in meat. Although all the connective tissue types are present in meat, fat (adipose tissue), bone, cartilage, and connective tissue predominate. The muscle and connective tissues are the major compositional components and contribute almost exclusively to the qualitative and quantitative characteristics of meat.

After the portions of the animal that are not generally consumed by humans are removed, there remains, on average, about 50 percent of the live weight of the animal on the carcass. Skeletal muscles comprise the main part of the weight of the meat carcass. Skeletal muscle, which is primarily what we eat, contains approximately 75 percent water, 20 percent protein, 4 percent fat, and 1 percent minerals and vitamins (evaluated as ash).

COMPOSITION OF MEAT

RAW COMPOSITION	Lean Beef	Lean Pork	Chicken Breast
Protein	20–22%	20%	20%
Fat	4–8%	2–7%	1–3%
Moisture (H_2O)	70%	75	70
Carbohydrates (CHO)	0%	0%	0%
Ash (vitamins/minerals)	1%	1%	1%

Raw meat usually does not contain any carbohydrates or contains a negligible amount, due to recovery from rigor mortis, whereby glycogen in the muscle is used up.

MUSCLE COMPOSITION

As noted above, there are three major constituents of meat: water, protein, and fat. Protein can be broken down broadly into three types:

1. Proteins that are soluble in water are called sarcoplasmic proteins and represent 33 percent of the protein in the muscle. We see these as "blood" in a package before cooking.
2. Proteins that are soluble in concentrated salt solutions, called myofibrillar proteins, make up 56 percent of the total protein. These are the proteins of greatest interest for binding and water holding.
3. Connective tissue proteins, composed mostly of collagen, are insoluble and represent 12 percent of proteins.

SUMMARY TABLE:

Type of Protein in Meat Muscle	Soluble In	Percent of Total Muscle Protein
Blood (sarcoplasmic)	Water	33%
Muscle fibres (myofibrillar)	Salt	56%
Connective tissue (collagen/elastin)	Not soluble	12%

MUSCLE STRUCTURE

Muscles are a fairly complex network of structures held together by a thin layer of connective tissues called epimysium. The muscles themselves consist of bundles of muscle fibers, and each bundle is surrounded by another layer of connective tissue called perimysium. Within those bundles the muscle fibers themselves (myofibers) are surrounded by their own layer of connective tissues called endomysial collagen as well as a layer of reticular fibers.

Myofibers are composed of myofibrils, and these structures are what give some cuts of meat a striated appearance, which means the muscle tissue is marked by dark and light diagonal bands. While collagen is present in all tissues and organs, the muscles of limbs (e.g., legs, shanks) that are more active contain more collagen. These muscles are less tender than muscles that are less active, such as those from the back (e.g., tenderloin). If collagen fibres are heated in water for a short time at 62°C, they contract and shorten, resulting in tough meat. However, when subjected to prolonged heating in water at temperatures above 70°C, these fibres solubilize (i.e., dissolve in water), and the result is tender meat. When you're planning a meal, it helps to know that the more collagen you have in the meat the more heat and time you'll need to cook the meat and still ensure a juicy and succulent dish.

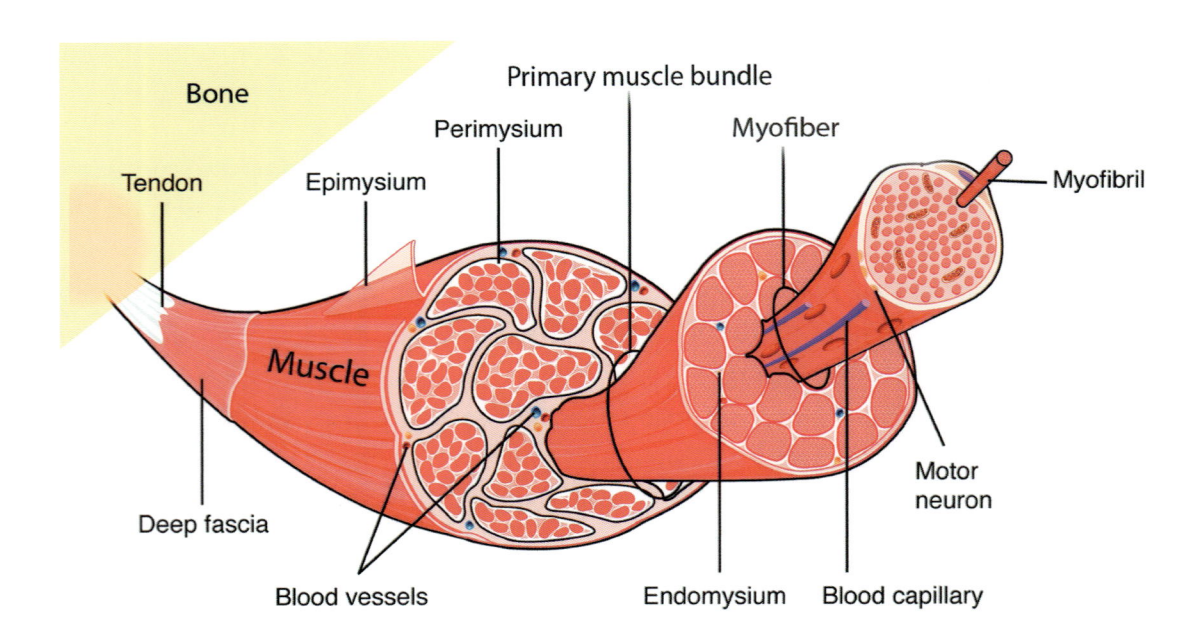

Schematic of skeletal muscle structure: An overview of a cross-sectional area of muscle, highlighting the muscle fiber structure

Fatty tissues have a different connective matrix that is not as strong as other connective tissues within a muscle. Consequently, this kind of connective tissue is looser and basically fills space between the meat bundles and other connective tissues. Fat droplets are stored in cells either intramuscularly between muscle bundles (this creates the marbling in some cuts of meat) or extra muscularly in the spaces between the muscle bundles, as well as subcutaneously between skin and muscles (e.g., backfat). Fatty tissues can be graded as hard and soft (e.g., backfat, jowl, and brisket) depending mainly on their connective tissue content. The important thing is, when cooked, these fat deposits melt and fill the space between the bundles, adding moisture and flavour that results in a delicious and succulent dish.

176

10. Turning Up the Heat on Meat

MY STORY: TOUTING TWO TS

"Time and temperature" is what I asked everyone at a recent leadership summit to remember. Time and temperature are what I tout in the restaurant to my chef apprentices. I grew up with the low-and-slow method of cooking. Knowing how quickly bacteria can grow and persist, I always thought this practice was risky from a food-safety perspective. In school – I can't remember in which grade it was – the concept of total energy turned on a light bulb for me. Employing heat energy with molecules moving faster for shorter periods of time was more efficient, expeditious, safer, and equal to the number conceptually of molecules moving slowly for longer periods of time. Funnily enough, this does work! It's just math. I'm always looking for a way to make more food cook and chill more quickly and easily. At work, in the test lab or in the manufacturing plant, we always need to expand capacity and look for room and space where it doesn't exist, so this concept works in our favour: hot, humid, blast chill, and go! We apply the same principles at the restaurant where there is no low and slow, only hot and fast (and blast chill). At home I work out how fast I can cook a chicken, braise the cheapest cuts of beef, or grill a steak because the math just works. Heat energy is equal whether it's fast or slow.

THE MAGIC OF SCIENCE: HEAT TRANSFER

Change of state magic: The concept of heat transfer was introduced to me in elementary school when I learned the changing states of matter through water that is molecularly the same but can exist as gas in the form of steam, or as a solid, such as an ice cube. Frozen water in the form of an ice cube melts, then forms a liquid and finally turns to steam when energy is added. The concept of heat transfer to change the state of matter is represented by two of Newton's laws of thermodynamics.

The first law of thermodynamics states that energy is neither created nor destroyed. The second law states that spontaneous heat transfer always occurs from a region of high temperature to another region of lower temperature. It is often called diffusion, where the direct exchange of kinetic energy passes through the boundary between two systems to achieve equilibrium. Often, as the two systems approach equilibrium, the rate of temperature change slows, just like a rollercoaster coming to a stop unless more energy is added to the system. These concepts are highly pertinent for cooking safely, efficiently, expeditiously, and, interestingly, cheaply.

Humidity magic: Heat transfer is more efficient using water, moisture, or steam than through the use of air velocity. Heat and humidity can be manipulated to create quality and sensory attributes for texture, colour, and flavour development. Of course, the density of the meat is important in considering how long to cook it.

Cooling magic: Placing cold meat in an oven, it temporarily drops the oven temperature due to the principle of heat/cold exchange. The greater the temperature difference between the cold meat and the surrounding oven, the sooner equilibration is achieved. Similarly, a hot dish just taken out of the oven cools faster in the fridge or, better yet, the freezer, compared to letting it sit out on the counter. Since the temperature difference between the hot dish and the cold fridge is higher, the food cools faster.

Step cooking magic: When I was working with a food engineer at one point, we got into a debate about what's called "step cooking." Step cooking involves raising the temperature in the oven very slowly, adding more energy one degree at a time. We started at 60°C and increased the temperature to 61°C, then later to 62°C, and so on, one degree at a time until we had reached the heat requirement for the food product. While this process makes a superb food product, it takes forever because each time you raise the temperature, you have to wait for equilibrium to happen! As I said before, I'm always in a hurry and looking for ways to cook food quickly and safely.

COOKING MEAT

Cooking is a heating process, the main objectives of which are to produce palatable and safe food. There is a lot going on during the cooking process: harmful bacteria are inactivated; the muscles shrink and stiffen creating a firm texture; moisture is lost; collagen is broken down so that tough, less desirable cuts of meat are tenderized; the meat colour changes; and, if searing or grilling meat, browning occurs and wonderful aromas develop.

Tenderizing: Heat can cause both tenderization and toughening of meat. Specifically, the myofibrillar proteins react to heat, altering the meat's natural qualities. The proteins coagulate and tighten and lose its ability to hold water. Tender cuts do not need a long time to cook. The temperature at which protein hardening occurs is imprecise, but most research suggests this does not occur below 64°C. Heating above this temperature results in less tender and, in some cases, quite rigid, well-done pork chops! They need longer heating at higher temperatures to break down the intramuscular connective tissue for tenderization.

Tougher cuts of meat, such as cheaper cuts such as blade roasts and leg and shank meat, contain a lot of collagenous connective tissue, so they require a different heating approach. Typically, these cuts are from muscles that are always in use by the animal. When the muscle is surrounded by the collagen of the connective tissue, the collagen begins to shrink to ⅓ of its original size, starting at 56°C (133°F), and completes the process between 62°C and 70°C. At temperatures near 72°C –74°C, a rapid shrinking of collagen is followed by protein hardening and toughening that can be later tenderized with continued heat at these temperatures. Collagen melts, fat melts, and collagen allows fat to move around (i.e., translocate) inside the muscle, providing juiciness and tenderization of muscle meats.

Cooking methods: Cooking, in the broadest term, embodies at least six methods: boiling, baking, broiling, frying, roasting, and stewing. The method of applying heat energy is different for each of these processes and can be reframed when you know there are three ways heat is transmitted and understand the process that occurs when transmitting that heat. Learn how to apply these methods and their effects and choosing the right cooking method becomes easy.

There are three distinct modes of heat transmission associated with cooking. *Conduction* occurs when one region of a solid body is at a higher temperature than another. When this occurs, heat flows to the region of lower temperature. *Radiation* occurs when two bodies are separated in space at different temperatures. All bodies emit radiant heat continuously. The intensity of that radient heat is dependent on the temperature and surface of the body. *Convection* is the process of energy transmission in which a fluid or gas is heated by conduction, then moves to a colder region, where it gives off heat also by conduction, typically by forced air.

Water or steam cooking (conduction) employs water as a conductor of heat energy in the form of steam or heated water. This is the most effective method from a heat-transfer perspective. The food comes into direct contact with the moist heat. Poaching, boiling, stewing, and steaming are obvious cooking methods where meat or food is submersed in water or steam during the cooking process. Braising and pressure-cooking use moisture conduction in a sealed environment where a small amount of moisture and the meat's natural moisture create a hot, humid environment. Sous vide methodology is essentially water cooking but at more controlled temperatures.

Dry-heat cooking (radiation) is the most common cooking method, noted by the "Bake" setting on most ovens. We often use dry-heating radiation for baking, roasting, and broiling in an oven. The heat radiates from the walls of the oven to increase the temperature of the food to be cooked. It is a slower heat transfer, and the moisture loss is great compared to a moister cooking method. Typically, much higher temperatures are used in this cooking method. Barbeques, which tend to cook at high temperatures, also depends on radiation.

Air cooking (convection) is known mostly as convection cooking. Most ovens use radiation from the bottom element to cook food. A convection oven incorporates a fan to create wind velocity that moves hot air around the oven, surrounding the food so moisture wicks from the meat. This process quickly causing dehydration on the surface while creating a crust that seals in juices, aids cooking time, and accelerates browning.

Air cooking with humidity (convection, radiation, conduction) plays an important role in efficient cooking processes. The use of radiant heat, together with air velocity saturated with moisture, creates the most effective heat conductor of all. High humidity generally results in more rapid heat transfer, lower cooking losses, and softer exterior. To create humidity in your oven at home, place a bowl of water in the bottom of the oven and replenish as needed throughout the cooking time.

Deep frying (conduction) is like water cooking, where the energy from the hot oil is in direct contact with the food and cooks it more quickly than water cooking. Oil is a higher conductor of heat than water, so the heat transfer from the oil to food for cooking is very fast and highly efficient. Check out my Beer-Battered Cod (p. 165).

- Classic Pulled Pork
- Beef and Barley Stew
- Prime Rib
- Rabbit Stew en Croute
- Bœuf Bourguignon (Pot Roast with Red Wine Sauce)
- Pork Chops with Brandy Cream Sauce
- Osso Bucco

CLASSIC PULLED PORK

MAKES 6 SERVINGS

2.2 lbs (1 kg) boneless pork shoulder, blade, or butt
1 ½ tbsp smoked paprika
1 ½ tbsp brown sugar
1 tsp black pepper

3 tsp salt
1 tsp dry mustard
1 tsp chili powder
¼ tsp cayenne pepper

PREPARING THE MEAT
1. Ensure the rind is off the pork shoulder and expose the lean meat as much as possible. Slash the meat with a knife or cut it into equal-size pieces.
2. Prepare the spice rub by mixing all the dry ingredients together in a small bowl. Set the rub aside.
3. Pat the meat dry and evenly apply half the rub over the surface of the meat. Cover the meat and refrigerate for 8 hours.

BRAISING IN THE OVEN
1. Place the refrigerated meat in a large Dutch oven or roasting pan and rub the rest of the spice rub all over the meat.
2. Cover and seal the cooking dish tightly with foil.
3. Preheat the oven to 400°F.
4. Place the covered meat in the oven and cook for 1 hour. Turn the oven down to 350°F and cook for 2–4 hours more until the meat is tender and pulls apart with a fork. (Times will vary depending on the quantity of meat.)

INSTANT POT
1. Place the refrigerated meat in the Instant Pot.
2. Rub 3–4 tsp of the rub on top of the meat and add ¼ cup liquid (water or stock) to the bottom of the pot. Put the lid on and close.
3. Set the Instant Pot to meat/stew and cook for 55 minutes.
4. Remove the meat from the Instant Pot. Let the meat cool enough that you can handle it.

PULLING THE MEAT
1. Pull the meat apart with two forks, eliminating any big chunks of fat but incorporating the rest.
2. Mix the meat juices with the meat.
3. Taste to check the seasoning and season with salt as needed.
4. Serve the pulled meat on a toasted bun with a little BBQ sauce, dill pickles and coleslaw (see below)

> ***Rule of thumb:*** Pressure cookers and Instant Pots cook food in high humidity, high-temperature conditions like braising but add high pressure to increase heat energy. The food cooks much faster as a result.

COLESLAW

3 cups shredded green cabbage
½ cup mayonnaise
½ cup shredded carrot
3 tbsp lemon juice

½ cup shredded red cabbage
¼ tsp black pepper
¼ tsp ground celery seed
½ tsp salt

1. Mix all the ingredients together and serve immediately.
2. Refrigerate for up to 2 weeks.

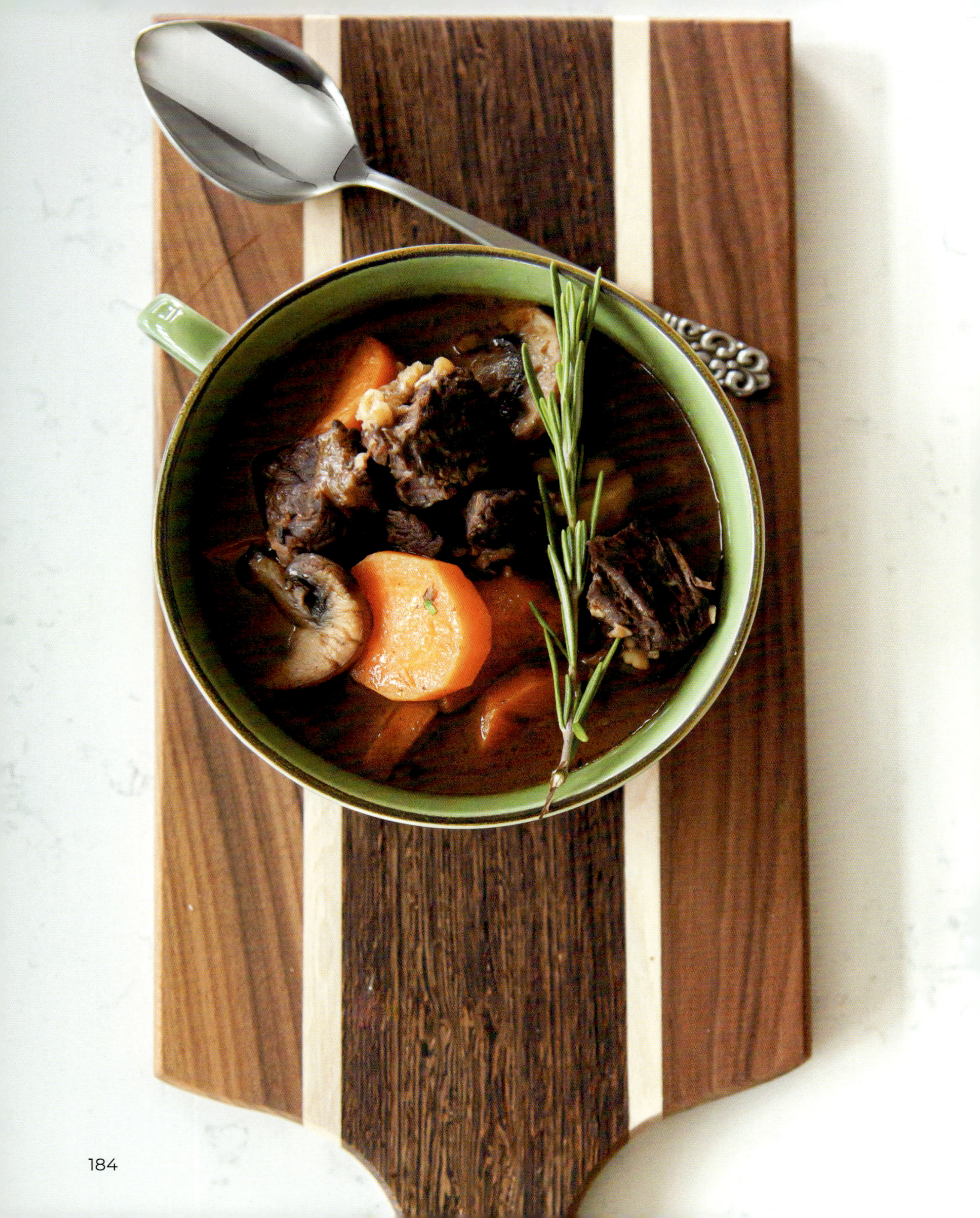

BEEF AND BARLEY STEW

MAKES 4–6 SERVINGS

This recipe is inspired by Martha Stewart's Beef and Barley Stew. I love the barley texture and flavour that develops with adding rosemary. This modified recipe really takes cheaper cuts of meat, gives them time to tenderize, and creates a flavourful, unexpected richness at any time of year.

2 tbsp olive oil
2 lb stewing beef, ½" cubes
Salt and pepper to taste
1 large onions, cut in ½" chunks
2 medium carrots, cut in ½" chunks
2 medium parsnips, cut in ½" chunks
8 oz mixed mushrooms; cremini, white, and chanterelles, thinly sliced

2 tsp minced garlic
2 tsp tomato paste
¾ cup dry red wine
3 cups beef stock (see p. 69)
Bouquet garni (rosemary, thyme, bay leaf)
¾ cup uncooked pot barley

1. Heat 1 tbsp oil in large stock pot or Dutch oven. Season the beef with salt and pepper. Brown the meat for 5–7 minutes in the hot oil, then remove the meat and set it aside.
2. Add and heat another 1 tbsp of oil. Sauté the carrots, onions, parsnips, and mushrooms until golden, 12–15 minutes. Season with salt. Remove the vegetables from the pot and set aside.
3. Add the garlic and tomato paste. Cook for 2 minutes.
4. Add the wine and deglaze the pot, scraping all the browned bits from the bottom.
5. Add the beef and vegetables back to the pot. Add the stock and water. Simmer for 1 ½ hours until the meat is tender.
6. Add the barley and the bouquet garni and cook for another hour.
7. Adjust seasoning and serve.

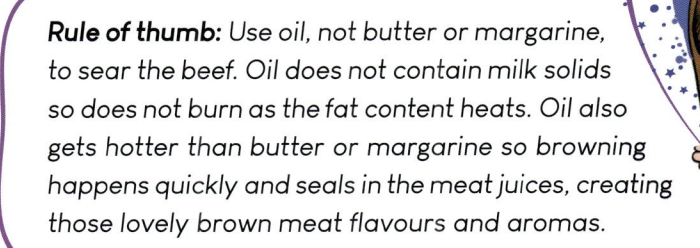

Rule of thumb: Use oil, not butter or margarine, to sear the beef. Oil does not contain milk solids so does not burn as the fat content heats. Oil also gets hotter than butter or margarine so browning happens quickly and seals in the meat juices, creating those lovely brown meat flavours and aromas.

PRIME RIB

Prime rib sounds daunting, but it is amazingly simple. The use of heat, both dry and moist, if used carefully, results in a perfect dinner without much work.

1 rib
1 tsp sea salt
½ tsp black pepper
2 tbsp olive oil

1. Season both sides of the prime rib with salt and pepper.
2. Heat a frying pan with 1 tbsp of olive oil until the oil begins to smoke.
3. Sear one side of the prime rib and then the other side.
4. Alternatively, cook the meat in a hot oven at 450°F to seal in the juices.
5. Then turn the oven down to 350°F and cook the rib until it reaches an internal temperature of 150°F (for a medium cooked rib).
6. Serve the drippings (i.e.. au jus) as is or use them to make a gravy (p. 89).

Start with bone-in, well-marbled beef. Bones don't add flavour, but they do regulate temperature, increasing the amount of tender, medium-rare beef you'll get in your finished roast.

Marbling is intramuscular fat that appears in a white, spider-web pattern within the meat. The more marbling, the richer and more tender your beef will be.

Rule of thumb: Most guides recommend a pound per person when you're shopping for prime rib. That amount is for very hungry eaters, or if you want leftovers. In reality, you'll most likely get away with three-quarters of a pound per person, or about one rib for every three people.

Season the meat well, and season it early if you've got time If you're able, it's best to season your prime rib with salt the day before, letting it sit on a rack in your fridge uncovered. Prime rib has plenty of flavour on its own. Just add a good heavy sprinkling of salt and pepper. Seasoning ahead of time allows time for the salt to penetrate and season the meat more deeply while also drying out the surface, which leads to better browning during roasting.

Rare, medium rare, or medium cook? Well-marbled prime rib is at its best when it's cooked to a minimum of medium rare, preferably medium. Rare is great for lean cuts such as a tenderloin that tend to get dry at higher temperatures. But for fatty, well-marbled cuts, you want to cook them at least to the point where the fat starts to soften and render, delivering flavour and juiciness. I've conducted my own blind taste test of beef cooked to various temperatures. all but one of a dozen tasters preferred prime rib cooked to medium rare, even the folks who initially claimed they liked their meat rare. Just a little more heat is a good thing for full flavour. Check out p. 7 for the cooking temperatures.

RABBIT STEW EN CROUTE

Any wild game meat is naturally dark in colour and contains lots of connective tissue, so it is rich with flavour but tough if not cooked well. Wild animals work their muscles for longer periods of time compared to their farm-raised counterparts. High, moist heat breaks down the myelin sheath covering the muscle bundles and the intermuscular connective tissue, getting deep into the muscle to melt the collagen and tenderize the meat.

MAKING PULLED RABBIT

MAKES 4 SERVINGS

Here are three different ways, employing different heat-energy delivery methods for cooking the rabbits.

STOVE (WATER COOKING—HIGH HUMIDITY)

1. Place four dressed rabbits into a large pot with water just covering the top of the rabbit.
2. Add ½ cup (25 percent) of beef stock to the water for flavour.
3. Bring the liquid to a boil for 1 ½–2 hours or until the meat pulls from the bones.
4. Let the rabbit and liquid cool slightly, strain, and pull the meat from the bones. Keep the strained liquid for the stew.

OVEN (BRAISING—HIGH HEAT, HIGH HUMIDITY)

1. Preheat the oven to 375°F.
2. Place four dressed rabbit into a roasting pan, Dutch oven, or deep baking dish. Season with salt and pepper, a dash of garlic powder, and a dash of onion powder.
3. Add ½ cup of beef stock to the bottom.
4. Seal the top of the cooking dish with a tight-fitting cover or foil.
5. Bake for 1 ½–2 hours, checking to see there is still some liquid at the bottom of the dish. If all the liquid has all been absorbed, add another ½ cup of water or stock to the bottom of the dish, recover and seal, then returning to the oven.
6. Check to see that the meat pulls easily from the bones. If not, continue cooking until it does.
7. Let the rabbit and liquid cool slightly, strain, and pull the meat from the bones. Keep the strained liquid for the stew.

INSTANT POT (PRESSURE—HIGH HEAT, HIGH HUMIDITY, HIGH PRESSURE)

1. Place four dressed rabbits into the Instant Pot.
2. Season the rabbit with salt and pepper, a dash of garlic powder, and a dash of onion powder.
3. Add ½ cup of beef stock to the bottom.
4. Put the cover on. Set the Instant Pot to the meat/stew setting.
5. Cook for 45 minutes to 1 hour.
6. Check to see if the meat pulls easily from the bones. If not, continue cooking for another 15 minutes on the manual setting.
7. Let the rabbit and liquid cool slightly, strain, and pull the meat from the bones. Keep the strained liquid for the stew.

Rule of thumb: Rabbits are very lean and small. One rabbit, farm-raised or wild, yields approximately ¼ lb of meat or one serving.

MAKING THE STEW EN CROUTE

MAKES 6–8 SERVINGS

1 lb cooked pulled rabbit (from 4 rabbits)
½ cup butter
¼ cup diced onions
½ cup flour
2 tsp minced garlic
3 ½ cups vegetable stock
2 cups beef stock
2 cups diced carrots
2 cups diced potatoes
3 tbsp fresh parsley, finely chopped

1 tsp fresh thyme
¾ tsp onion powder
¾ tsp garlic powder
½ tsp Mount Scio savoury (optional)
1 sheet premade frozen puff pastry, slightly thawed
Salt and pepper
1 egg
1 tbsp water

1. Melt the butter in a large pot. Add the garlic and onions, and sauté until the onions are soft. Allow the onions to caramelize and turn brown.
2. Add the flour and stir constantly as it cooks and creates a roux, approximately 2 minutes.
3. Add the beef stock slowly while whisking to ensure there are no lumps.
4. Add all the vegetable stock and then the carrots and potatoes.
5. Bring the mixture to a rolling boil, then simmer on medium low for approximately 30 minutes until the carrots and potatoes are cooked.
6. Add the pulled rabbit, herbs, and spices. Stir to mix thoroughly. Add salt and pepper to taste.
7. Portion in individual serving dishes or one baking dish, and cover with the puff pastry. Cut some slits for steam to escape. Brush the pastry with an egg wash made from 1 egg and 1 tablespoon water beaten together (see below).
8. Cook at 425°F for 10–15 minutes.

WHAT IS EGG WASH? WHY IS IT USED?

Brush an egg wash on the top of baked items to quickly develop a golden brown colour that doesn't naturally occur during the cooking time. Egg has natural ingredients such as proteins that combine with starches and sugars in pastry to create the golden brown colour quickly. An egg wash is easy to make. Just crack an egg in a bowl, add 1 tbsp of water, and whisk until the mixture is uniform and yellow. Brush on top of the pastry just before placing in the oven.

BOEUF BOURGUIGNON

MAKES 4–6 SERVINGS

½ lb (225 g) bacon, diced fine
1 medium onions, cut lengthwise into slices
250 g mushrooms, cut in half
2 lbs (1 kg) beef round roast, sirloin tip, or stewing beef, cut into large cubes
3 tbsp flour
3 cloves garlic
1 ½ cup beef broth
1 ½ cup full-bodied red wine
3 bay leaves
1 tsp dried thyme
½ tsp salt
½ tsp pepper

1. Preheat the oven to 350°F.
2. In a large Dutch oven or heavy stockpot, fry the bacon over medium heat until brown. Remove the bacon and set aside.
3. Sauté the onions and mushrooms in the bacon fat until just browned. Remove the vegetables and set them aside.
4. Brown the beef in the bacon fat, adding some vegetable oil if needed. Sprinkle the flour over the meat once it's brown.
5. Add the garlic, wine, broth, bay leaves, thyme, salt, and pepper. Cover the pot tightly and transfer to the oven for 45 minutes. Reduce the oven temperature to 325°F for an hour until the meat is tender and the sauce is somewhat rich in colour.
6. Add the bacon, onions, and mushrooms and cook for another half hour.
7. Add salt and pepper to taste and serve hot.

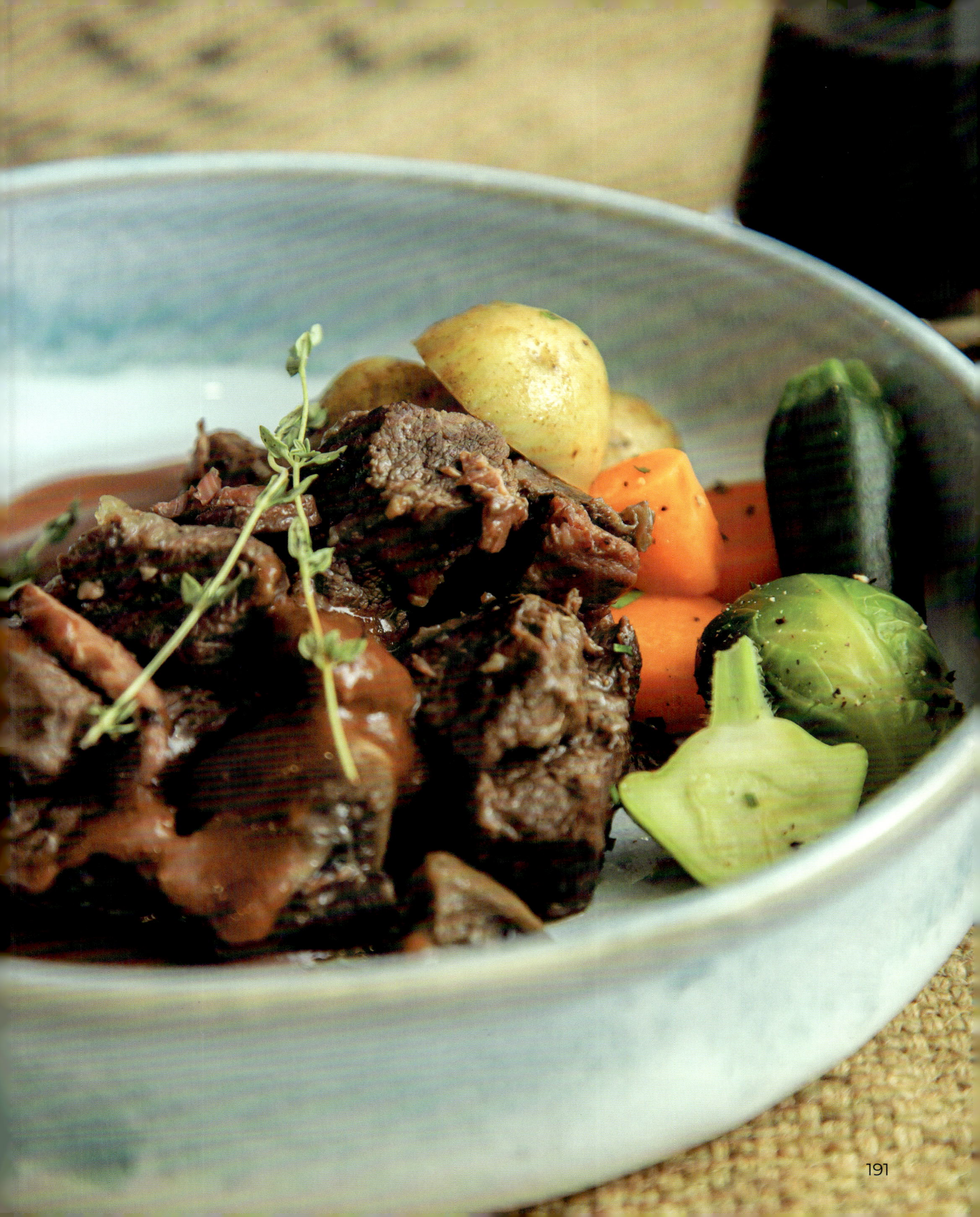

PORK CHOPS WITH BRANDY CREAM SAUCE

MAKES 4 SERVINGS

4 thick loin pork chops, bone in
Salt and pepper to taste
1 tbsp oil
1 tbsp butter
2 medium onions, sliced thin
2 tart apples, such as Granny Smith, peeled, cored, and sliced

3 tbsp brandy
1 ½ cups beef or vegetable stock
1 tbsp flour
A pinch grated nutmeg
¾ cup whipping cream

GARNISH:
2 more apples, cored, unpeeled, and sliced
2 tbsp butter
2 tbsp sugar

1. Generously season the pork chops with salt and pepper.
2. Heat the oil and butter in large frying pan and fry the pork chops over medium heat until they're brown on both sides. Remove and set aside.
3. Add the onions to the pan and cook them until they're soft but not brown.
4. Add the peeled, sliced apples and cook over high heat until both the onions and apples are golden.
5. Add the chops back to the pan and pour the brandy over them. Flambé the chops. (Be careful!)
6. Stir the flour into the juices. Add the stock and nutmeg and bring to a boil.
7. Cook and simmer everything on the top of the stove for 1 hour (or cook in the oven at 350°F for 1 ½ hours). If the sauce thickens, add some more stock.
8. Remove the chops and put the sauce through a sieve to extract the apple pulp and onions.
9. Return the sauce to the pan.
10. Add the cream to the sauce and bring to a boil.
11. Turn down the heat and reduce the sauce until it is thick enough to coat the back of a spoon.
12. Spoon the sauce over the chops and use the cooked apple slices to garnish them.
13. Sliced Apple Garnish: Heat the 2 tablespoons butter in the frying pan. Dip each apple slice in sugar. Cook the slices, sugar side down, in the butter for 4–5 minutes so the sugar caramelizes. Sprinkle more sugar on the uncooked side of the apple slice, turn, and brown the second side.

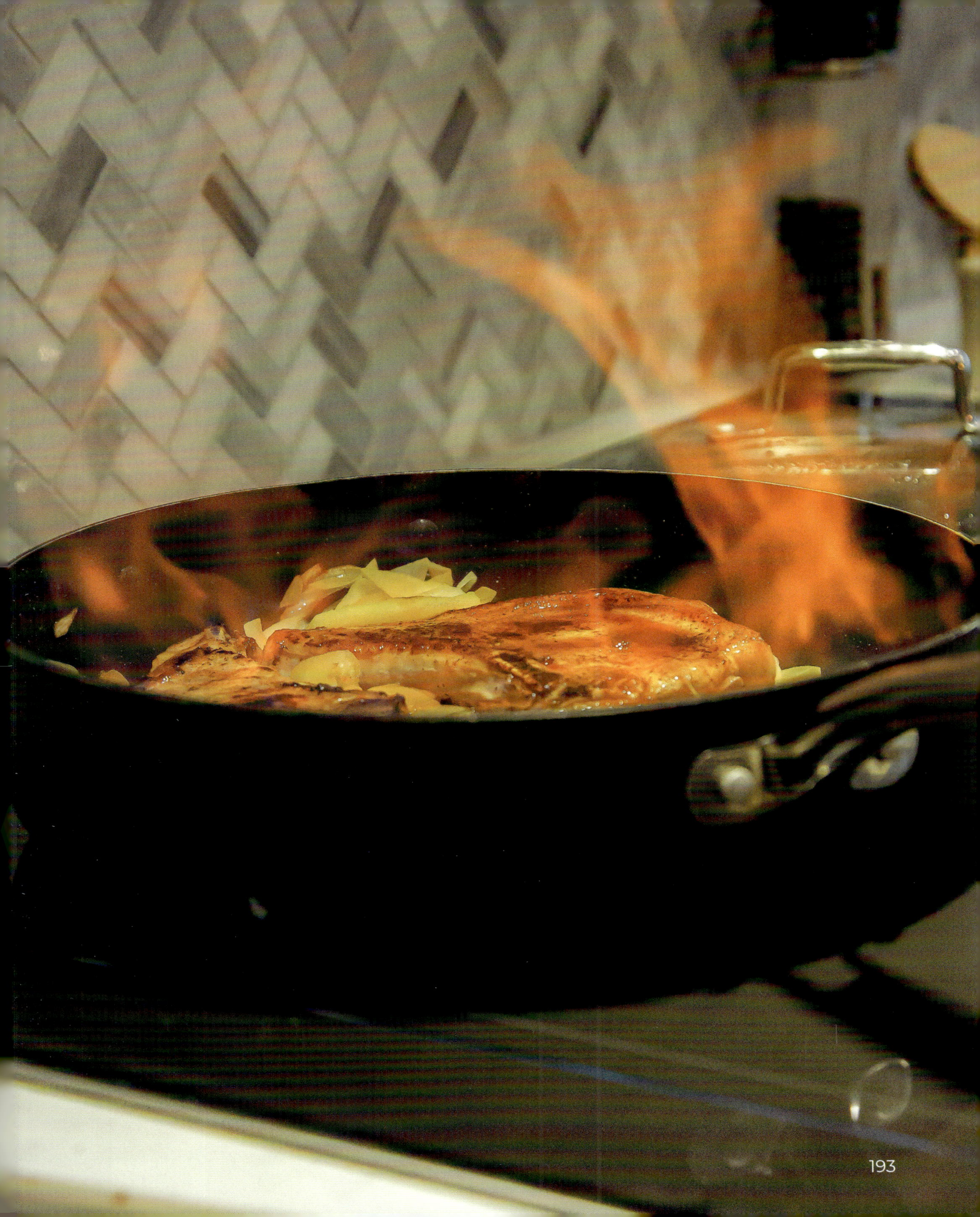

OSSO BUCCO

MAKES 4 SERVINGS

4 veal shanks slices (or pork shanks) about 1 ½ inches (5cm) thick
1 tbsp flour for dredging
2 tbsp olive oil
1 medium onion, chopped
1 medium carrot, chopped
2 stalks celery, chopped
4 cloves garlic, minced

1 19-oz can (540 ml) diced tomatoes
1 tbsp tomato paste
1 cup beef broth
1 ½ cups white wine
1 bay leaf
1 tsp dried thyme
Salt and pepper

Here are two different ways, employing different heat-energy delivery methods for cooking the veal (or pork):

OVEN (BRAISING—HIGH HEAT, HIGH HUMIDITY)

1. Preheat the oven to 350°F.
2. In a large Dutch oven, heat the oil over medium heat until hot. Generously season the flour with salt and pepper and dredge the shanks in the seasoned flour. Brown both sides of the shanks in the oil. Set aside.
3. In the same pan, cook the onion, carrot, celery, and garlic in oil about 5 minutes, stirring occasionally, until tender.
4. Deglaze the pot with white wine. Add the remaining ingredients and bring to a boil. Add the meat and return to a boil.
5. Cover the pot and transfer it to the oven for 45 minutes. Reduce the oven temperature to 325°F for an hour until the meat is tender and the sauce is somewhat caramelized in colour.

INSTANT POT (PRESSURE—HIGH HEAT, HIGH HUMIDITY, HIGH PRESSURE)

1. In the Instant Pot, press Sauté and heat the oil until hot. Generously season the flour with salt and pepper and dredge the shanks in the flour. Brown both sides of the shanks in the oil. Take the shanks out and set aside.
2. Add another tbsp of oil to the pot. Sauté the onion, carrot, celery, and garlic about 5 minutes, stirring occasionally, until tender.
3. Deglaze the pot with white wine. Add the remaining ingredients and bring them to a boil. Add the meat and return the pot to a boil.
4. Seal the cover of the pot and set to manual and cook on normal for 30 minutes.
5. Then set to slow cook for 2 hours until the meat is tender and the sauce is somewhat caramelized in colour.

ONCE THE MEAT IS COOKED:

6. Separate the bones and meat, and take out all the fat.
7. Strain out any solid bits from the sauce and retain the smooth velvet sauce. Place in a saucepan and simmer.
8. Adjust the seasoning as needed. Serve the dish with the bone, meat, and sauce complemented with garlic smashed potatoes (see p. 104) and asparagus, lemon and almonds (see p. 118) and gremolata.

HOW TO MAKE A GREMOLATA
This is a wonderful accompaniment for the Osso Bucco meat. Finely grate the zest of 2 lemons and add ¼ cup finely chopped parsley and ¼ cup finely chopped garlic cloves. Combine everything in a bowl. Spread the gremolata over the meat just as you're ready to serve the dish.

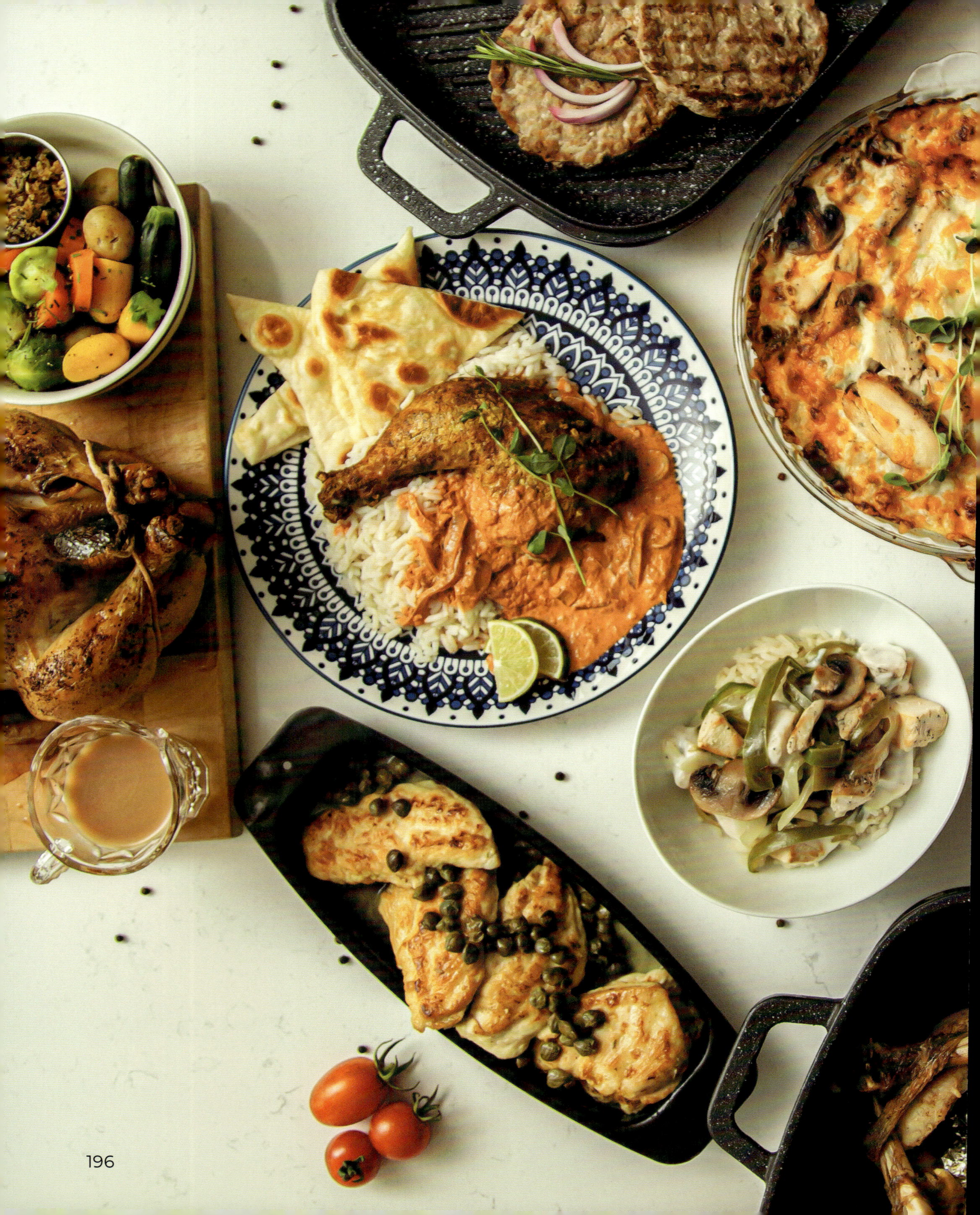

11. Bye Bye, Dry Birdie

MY STORY: BURGER MAGIC

A hamburger is delicious and easy to prepare, and most people eat hamburgers. It's a sure bet for a barbeque with family and friends. Everyone has a special burger recipe. But as soon as I announce my recipe does not include an egg or breadcrumbs, and I add water, there is disbelief. There is always a section in my food product knowledge seminars about burgers. My simple, clean, super-juicy burger recipe illustrates a fundamental of the science of meat and brings science to life in front of your eyes. In the Sofina Foods Executive Process Mastery training, the burger recipe is foundational. Two operations vice presidents fancied themselves barbeque experts. They took up the challenge to conduct my burger experiment on the weekend. After following my recipe and my process, and after sampling their product, they arrived ahead of me on Monday morning smiling and animated with a sample of their prized burgers. They were excited about how deliciously juicy the burgers were, how they cooked without fire flaring up or shrinking into small pucks. My hamburger recipe links a science concept to food so clearly. Once tried, the barbeque experts are delighted to have solved all their burger woes forever. Many a convert greets me time and time again, impatient to share their barbeque successes. The excitement is created by knowing the science, knowing why it happens, and being able to apply it themselves. The inspiration for this cookbook is to share widely how to connect science to everyday food, such as a super-juicy burger.

THE MAGIC OF SCIENCE: WATER-HOLDING CAPABILITIES

In Chapter 9, the magic in the meat is the inherent moisture in meat and the functional properties of the meat muscles in their abilities to emulsify fat and hold moisture. The overarching concept, simply put, centres around water management—addition or subtraction—using the science of meat, heat transfer, and emulsification for tasty, juicy meat. Effective water holding begins with liberating the functional myofibrillar proteins - actin and myosin, specifically - from the meat muscle. Doing this make the meat muscle available and effective for water holding, emulsification, and texture development and binding. A small amount of salt acting to extract actin and myosin from the muscle, bringing them to the surface, creates a sticky, gluey texture. This stickiness indicates the proteins are ready and available to hold water, emulsify fat, and bind to other meat. As a rule of thumb, consider that 1 lb (454 g) of lean meat holds 20 percent of its weight (90 g or ⅓ cup) in water. Lean meat has minimal fat content to contribute to a juicy and tender meat product. Evaporation of some of the inherent moisture happens during the cooking process, so adding moisture such as water to the meat replenishes what's lost during the cooking process and enhances succulence. Water provides an additional benefit in that flavours such as herbs and other inclusions are better incorporated through mixing or infused into the muscle by migrating through the water.

The extracted sticky meat proteins also aid in emulsifying fat particles by coating and stabilizing fat globules within a protein-water matrix, similar to making mayonnaise. During the cooking process, these extracted proteins coagulate while holding the moisture, adhering to other proteins, and surrounding fat molecules to form a coherent matrix that is important for texture, succulence, and fat-retention during the heating process. This process of binding small and large meat particles is important for holding the meat product together. The process of holding moisture and trapping fat molecules is key to minimizing cooking losses. As well, controlling the temperatures during preparation and cooking is important to creating a desired juicy, flavourful meal such as burgers, chicken ballotine, or roast chicken.

In this chapter, the recipes and methods in this chapter let you make super juicy flavourful chicken and turkey dishes.

- Super-Juicy Burgers
- Stuffed Chicken Breast Ballotine
- Butter Chicken
- Chicken Piccata
- 15-Minute Chicken A La King
- One-Hour Roast Chicken Dinner
- Smoked Paprika and Lemon Garlic Flat Spatchcock Chicken
- Maple Sage Anytime Turkey
- Leftover Turkey Tetrazzini

Food is the ingredient that binds us together.
–Unknown

SUPER-JUICY BURGERS

MAKES 4 SERVINGS

This is the famous burger recipe. Yes, it takes just three ingredients and the magic of science.

1 lb (454g) lean or extra-lean ground meat (chicken, turkey, pork, beef), kept cold
1 tsp salt
⅓ cup cold water

SEASONINGS:
Pepper, garlic or garlic powder, finely diced onions or onion powder, chili powder

1. Put the cold ground meat in a bowl. Add the salt directly to the meat.
2. Work the meat with a spoon or your hands until it is tacky.
3. Add the water and work the meat until the water is absorbed.
4. Make four ¼ lb patties. Store them in the fridge overnight for juicier burgers.
5. Place the patties on a hot grill or griddle.
6. Flip the patties when the underside is easy to lift off. Flip one more time for the final sear.
7. Cook the burgers completely to 160°F (71°C).

Rule of thumb: Use ice cold water. The meat proteins are best extracted at cold temperatures around 0°C. When you keep the burgers overnight in the fridge they will absorb the water and hold on it more tightly. The burgers will not shrink or flame on the grill.

For more on making juicy burgers, check out the "How To" video on YouTube.

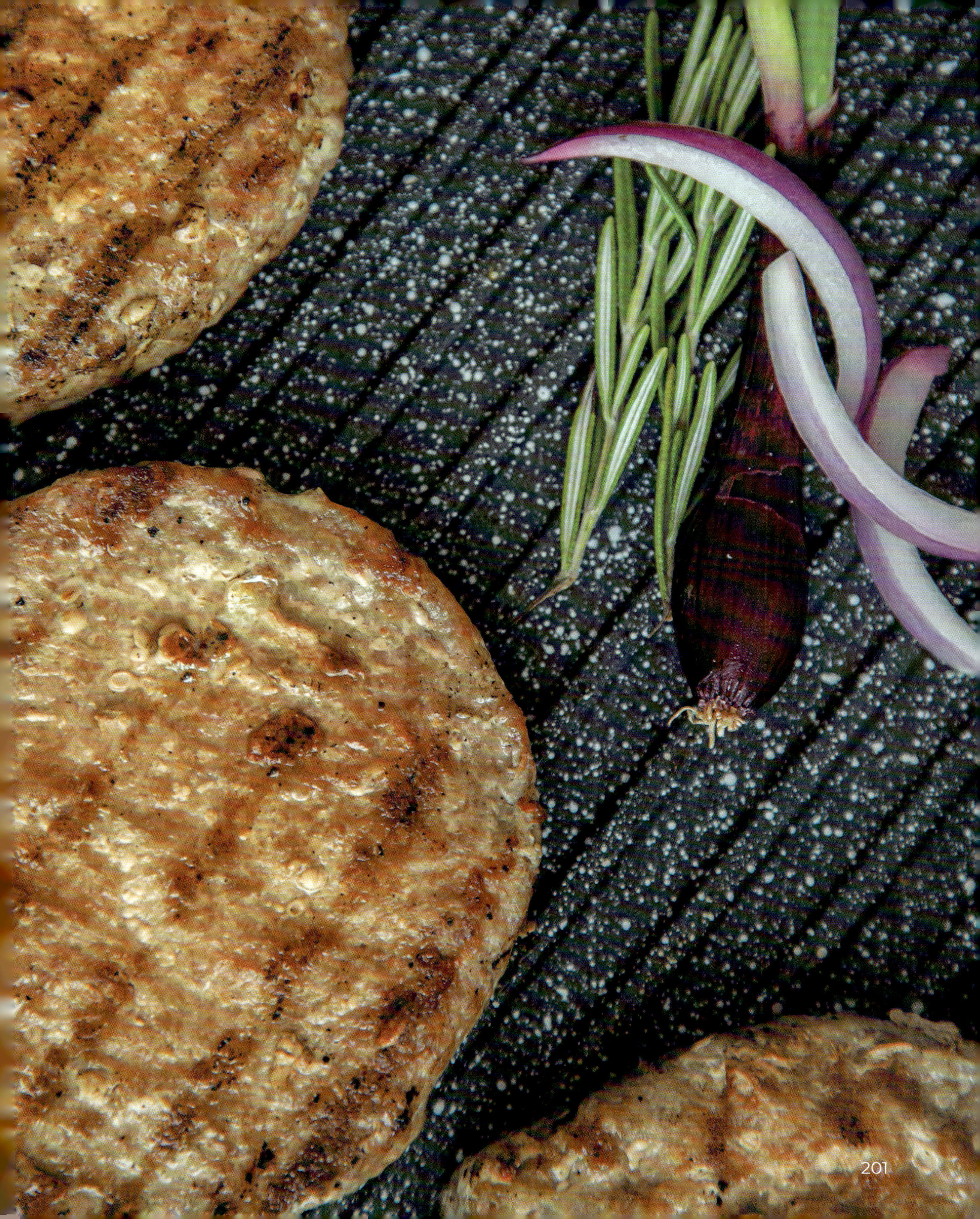

STUFFED CHICKEN BREAST BALLOTINE

MAKES ABOUT 4 SERVINGS

This is the *best* stuffed chicken breast! *Ballotine* comes from the French *balle en paquet*. Typically, it is a stuffed thigh or leg of chicken, but here we upgrade to a boneless chicken breast, with or without the skin. Stuff the chicken breast with onion pecan stuffing and drizzle with a velouté (see p. 89). You will be the chef in your own kitchen!

4 boneless chicken breasts (tenderloins removed), skin on or skinless
1 tsp salt
1 tbsp oil for searing

ONION PECAN STUFFING
3 tbsp onion, finely chopped
¼ cup soft breadcrumbs from real bread, not dried, commercial breadcrumbs
¼ cup pecans, toasted and rough chopped
1 clove garlic, finely chopped or minced
1 tsp dried parsley (or 1 tbsp chopped fresh parsley)
3 tbsp melted butter
Salt and pepper to taste

> Don't dry out the breadcrumbs for the stuffing or skimp on the butter. If you do, the chicken will offer its moisture to the stuffing as per Newton's law, and the chicken will be less moist.

1. Preheat the oven to 400°F.
2. Place the chicken breasts skin-side down on a cutting board.
3. Make multiple small cuts in the side of each breast to slice them open in a butterfly format, taking care to keep the breasts in one thinner breast piece.
4. Sprinkle the open meat with salt.
5. Put some plastic wrap over top of the breast and press or pound the meat with a rolling pin or mallet. The surface should be sticky, and the meat should be thinner and more uniform in thickness.
6. Prepare all the stuffing ingredients as indicated above. Mix together. Season to taste.
7. Add a couple of tablespoons of the stuffing to each of the prepared flattened chicken breasts, and roll the breast up.
8. Lay the stuffed chicken breasts seamside down on a tray. Put the tray in the fridge.

9. Let the stuffed chicken rest for 15 minutes in the fridge.
10. In an ovenproof pan, sear the chicken breasts in 1 tablespoon of oil, browning the outside only.
11. Finish cooking the breasts in the oven for about 10–15 minutes to an internal temperature to 160°F (70°C).
12. Let the chicken rest for a few minutes before cutting and serving with a little velouté (see p. 89).

For more on onion pecan stuffing, check out the "How To" video on YouTube.

BUTTER CHICKEN

This butter chicken recipe includes the whole thing. It is worth the effort. Use high heat for the chicken and cook the chicken legs at a little higher heat still so the dark meat is no longer pink and the juices are clear. The naan bread is a great complement and very fast to execute.

PERFECT BASMATI RICE
MAKES 4 SERVINGS

1 cup basmati rice
1 ½ cups water
1 ½ tbsp butter
½ tsp salt

1. Boil the water, stir in the salt and butter, then add the rice. Wait until the water boils again and stir the rice in the water.
2. Turn the heat to low, cover the pot, and let the rice absorb the water for 15 minutes. Do not stir the rice while it is as it is hydrating. Fluff and serve the rice with Butter Chicken or Chicken à la King.

BUTTER CHICKEN SAUCE
MAKES 4 SERVINGS

3 tbsp olive oil
1 tbsp butter
1 large onion, sliced
1 ½ tbsp garlic, minced (about 3-4 cloves)
1 tbsp ginger, finely grated (or 1 tsp ground ginger)
1 ½ tsp ground cumin
1 ½ tsp garam masala
1 tsp ground coriander
3 cups canned tomatoes, crushed or pureed in blender
1 tsp red chili powder
1 ¼ tsp salt
1 cup 35% fat cream
2 tsp sugar

1. Heat the oil and butter together in a large frying pan. Sauté the onions until translucent.
2. Add the ginger, garlic, and spices. Stir in and let cook for 1–2 minutes.
3. Add the tomatoes and simmer for 15 minutes or longer. Taste and adjust the salt and sugar.
4. Set aside until the chicken is ready.
5. Add the cream and heat the sauce before serving.

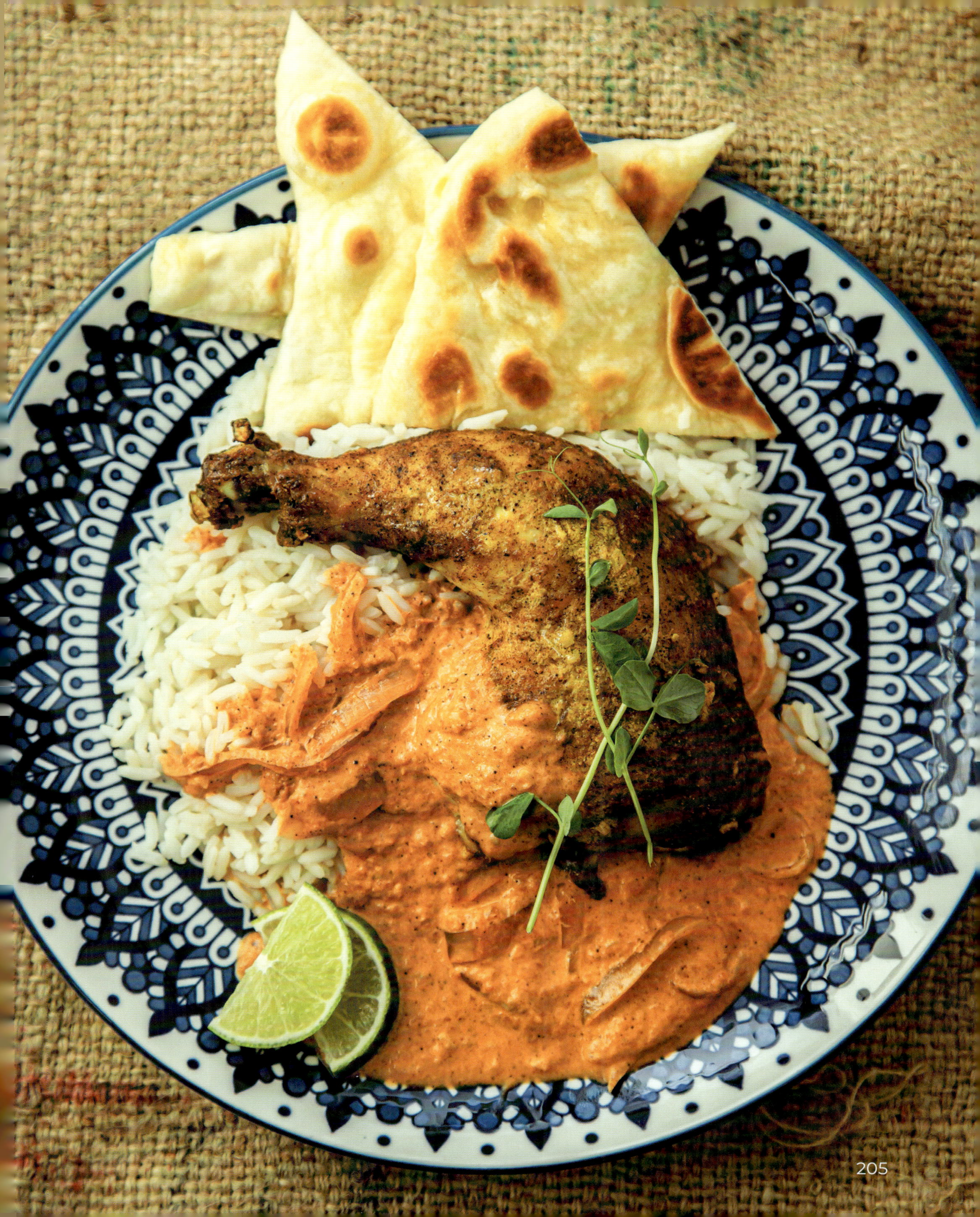

NAAN BREAD
MAKES 4 SERVINGS

3 cups flour
1 tbsp sugar
1 tsp salt
1 tbsp baking powder

½ cup water
½ cup whole milk
1 egg

1. Mix together the flour, sugar, salt, and baking powder. Set aside.
2. Whisk together the water, milk, and egg until combined.
3. Add the milk-egg mixture to the flour mixture slowly until it is all mixed in and forms a dough.
4. Turn the dough out onto a floured surface and work the dough, kneading until it is a smooth ball.
5. Cut the dough in half. Wrap each portion in plastic wrap and refrigerate to rest for 20 minutes.
6. Remove the plastic and roll out the dough very thin.
7. Heat oil in a hot frying pan. Place the dough on the hot pan and watch it rise. Flip the fried dough over and brown the other side.
8. Butter the naan and cut to serve with the Butter Chicken.

BUTTER CHICKEN
MAKES 4 SERVINGS

4 chicken legs, fresh or thawed
½ cup plain yogurt
1 ½ tbsp minced garlic
1 tbsp finely grate ginger (or 1tsp ground ginger)

2 tsp garam masala
1 tsp turmeric
1 tsp ground cumin
1 tsp red chili powder
1 tsp salt

1. Preheat the oven to 375°F. Mix all the ingredients together and slather on the chicken legs, even the underside. For more flavour development, marinate the chicken legs overnight in the fridge or 2–3 hours before cooking.
2. Cook the chicken legs for 35–45 minutes to an internal temperature of 175°F or until the juices run clear.
3. Serve with perfect basmati rice, butter chicken sauce, and some naan bread.

CHICKEN PICCATA

MAKES 4 SERVINGS

Thin butterflied chicken breasts can be cooked very fast and result in a super juicy, tasty dish. Add a little lemon acidity to aid in breaking the intramuscular connective tissue. It's fast and delicious.

4 skinless, boneless chicken breasts, cut butterflied
Salt and black pepper to taste
4 tbsp flour for dredging
¾ cup butter
½ cup vegetable oil
⅔ cup lemon juice
1 cup chicken stock
½ cup brined capers, rinsed
⅔ cup fresh parsley, chopped

1. Season the thin butterflied chicken breasts with salt and pepper. Dredge the chicken in flour and shake off the excess.
2. In a large frying pan over medium-high heat, melt 2 tbsp of butter with 3 tbsp olive oil. When the butter and oil start to sizzle, add two pieces of chicken and cook for 3 minutes.
3. When the chicken is browned, flip the breast and cook the other side for 3 minutes. Remove and transfer to a plate.
4. Melt 2 more tbsp butter and add another 2 tbsp olive oil. When the butter and oil start to sizzle, add the other 2 pieces of chicken and brown both sides in the same manner.
5. Remove the pan from the heat and add the chicken to the plate.
6. Pour the lemon juice, stock, and capers into the pan you used to brown the chicken breasts in. Return the pan to the stove and bring the contents to a boil, scraping up the brown bits of the cooked chicken from the pan for extra flavour. Check for seasoning.
7. Return all the chicken to the pan and simmer for 5 minutes.
8. Place the chicken on the platter to serve.
9. Add remaining 2 tbsp butter to the sauce and whisk vigorously.
10. Pour the sauce over chicken and garnish with parsley.

15-MINUTE CHICKEN À LA KING

MAKES 4 SERVINGS

This is one of my go-to, fast, delicious, and nutritious meals.

1 medium green pepper, cored and sliced thin
1 medium onion, peeled and sliced thin
3 tbsp butter
4 large button mushrooms, sliced

1 ½ cup thick béchamel sauce (see p. 84)
½ cup water
4 boneless, skinless chicken breasts, diced
Salt and pepper

1. Heat a large frying pan and add 1 tbsp of butter. Add the diced chicken to the hot pan.
2. Sear the chicken, turning the pieces so they get golden brown on the sides, sealing in the juices. The chicken does not have to be fully cooked, just brown on the outside. Remove and set aside.
3. Add another tablespoon of butter to the pan. Add the green peppers and onions. Cook for 3–4 minutes until they are soft.
4. Move the onions and green peppers to the side of the frying pan so they can continue to cook.
5. Add the sliced mushrooms, and sauté them until they're brown on each side. Mix the mushrooms with the green peppers and onions.
6. Stir in the béchamel sauce. Add water particularly if the sauce is too thick. Add the chicken and simmer until the chicken is completely cooked reaching an internal temperature of 160°F (71°C).
7. Serve on a bed of smashed garlic potatoes (see p. 104)

Rule of thumb: If you don't have the béchamel sauce made ahead of time, go ahead and shamelessly use a store-bought can of cream of chicken soup...Its base is a bechamel sauce. Using the soup works well especially if you're in a hurry.

Substitute leftover chicken for the chicken breast, skip the cooking step and let the chicken get hot in the sauce.

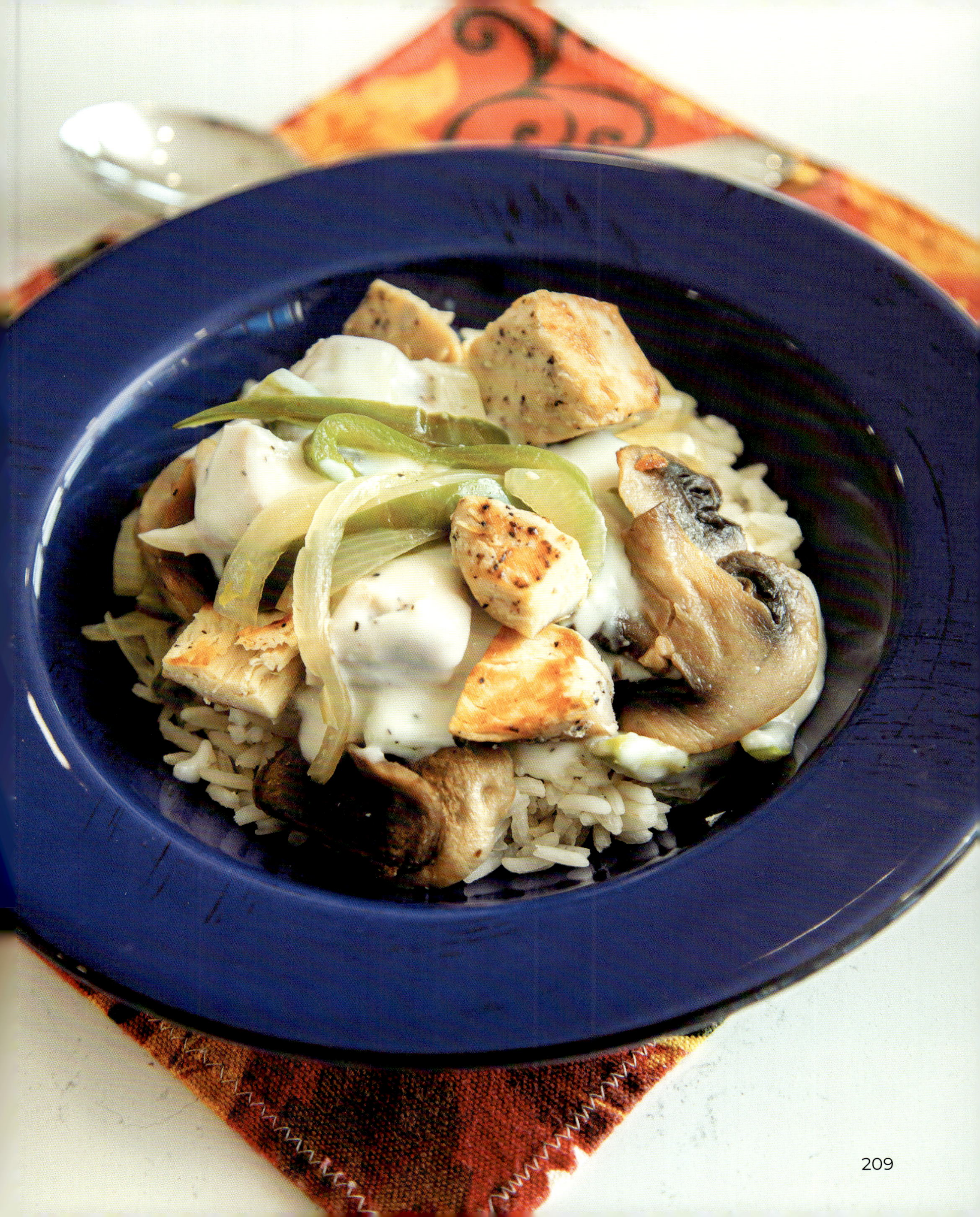

ONE-HOUR ROAST-CHICKEN DINNER

MAKES 4 SERVINGS

I posted a "How To" a video on YouTube that shows how to cook a chicken in one hour! This may seem frightening to most, especially if you're used to your Sunday dinners taking hours to cook. But I can show you just how to speed it along. Watch this video, and I'll show you exactly what to do!

1 kg chicken (2.2 lb chicken)
Salt and pepper to taste
String
Aluminum foil

piece of apple
peeled onion cut in quarters,
your choice of herbs such as
sage, parsley, rosemary.

1. Rinse the chicken inside with cold water. Add some apple and onion and herbs to the cavity.
2. Stuff the chicken with a balled-up piece of foil (to keep the chicken moist, place the foil ball at the opening of the chicken's cavity so it is sealing the cavity, like the entrance to a cave).
3. Preheat the oven to 400°F.
4. Wrap the string around the chicken to secure the legs and wings to the chicken's body. Doing this allows the heat to transfer more evenly so all parts of the chicken are cooked at the same time. That eliminates the problem of wings burning while the inner meat is still not cooked.
5. Put the chicken in a shallow dish. Cook for 20 minutes at 400°F. The outside of the chicken will be sealed.
6. Reduce the temperature to 350°F and continue to cook the chicken until its internal temperature at the thigh is 175°F (80°C) or the juices run clear, about 40 minutes.

7. Remove the chicken from the oven and put a little foil over the top of the chicken to create a tent that minimizes moisture evaporation and aids in heat retention.
8. Let the chicken sit to equilibrate or "rest" for 15 minutes.
9. Carve the chicken and serve with traditional down-home stuffing and a velouté (see p. 89).

Rule of thumb: Temperature change happens quickly at 450°F without drying the chicken and then equilibrates more slowly when the temperature is 350°F. If you would like to use convection heating, the moisture moves from high concentration to low concentration. The moisture inside the chicken moves quickly to the surface and is whisked away until the surface temperature reaches a protein denaturation temperature, around 144°F(65°C), at which time the moisture is trapped inside the chicken, creating its juiciness. The faster you can get to this point, with or without the convection setting, the less moisture escapes, resulting in super juicy, fast chicken. Bye-bye, dry birdie!

TRADITIONAL DOWN-HOME STUFFING

(also on p. 167)

2 cups fine breadcrumbs
½ cup finely chopped onions
½ tsp salt
¼ tsp black pepper
¼ cup melted butter
2 tbsp Mount Scio savoury

1. Preheat the oven to 350°F.
2. Add all the ingredients to a bowl and mix until combined.
3. Transfer the mixture to an ovenproof dish with a cover.
4. Cook for 30 minutes, until the onions are soft and the breadcrumbs are a little brown.

For more on cooking a whole chicken in an hour, check out the "How To" video on YouTube.

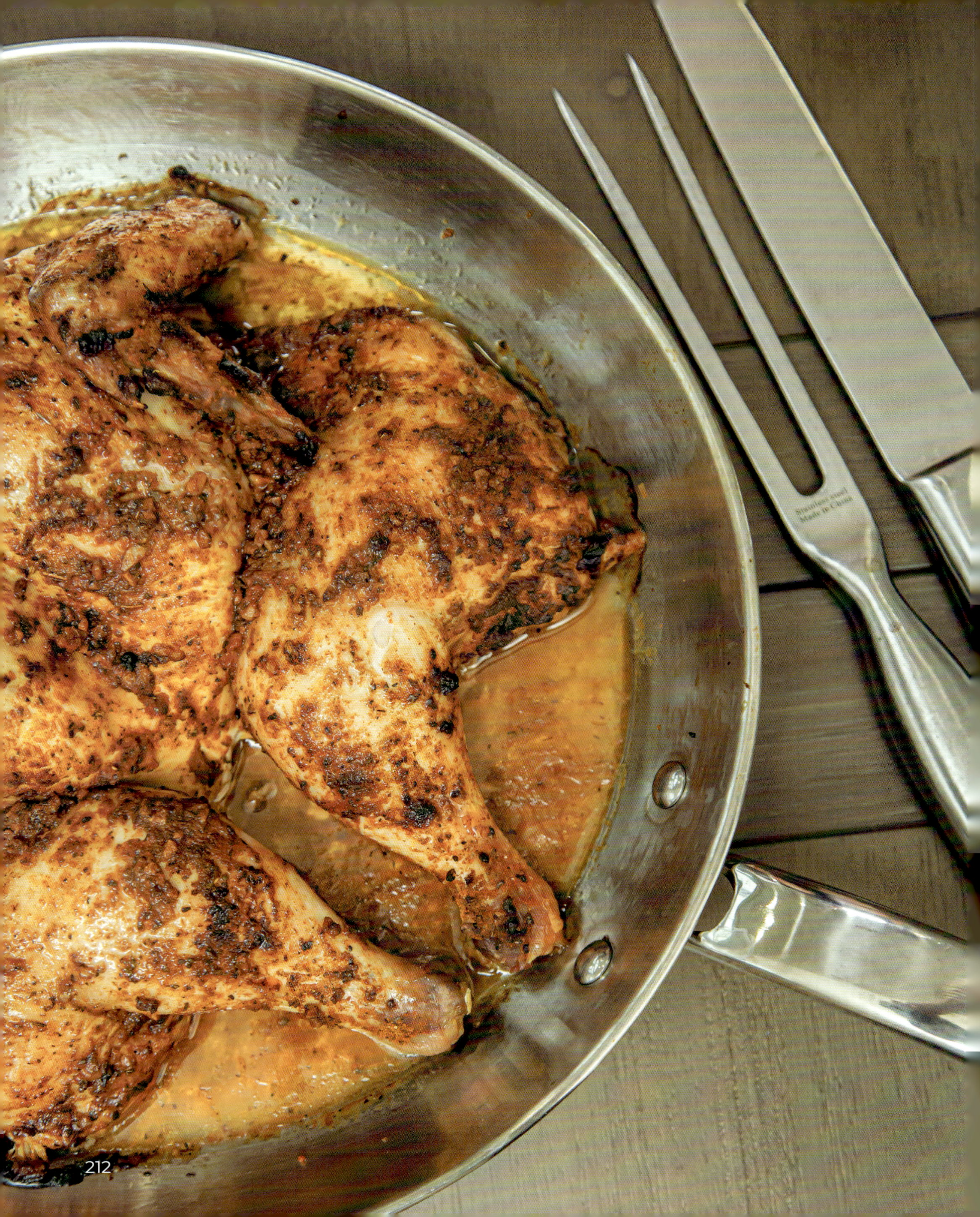
Stainless steel
Made in China

SMOKED PAPRIKA LEMON GARLIC FLAT SPATCHCOCK CHICKEN

MAKES 4 SERVINGS

1 whole chicken, weighing about 3 pounds, fresh or thawed

FOR THE RUB
1 tsp minced garlic
1 tbsp salt
1 tbsp smoked paprika
Juice from 1 lemon
1 tsp Italian seasoning
½ tsp freshly ground black pepper

Make your own Italian seasoning with equal parts basil, thyme, and oregano

1. Preheat the oven to 400°F.
2. Use scissors or a sharp knife to cut down the middle of the back of the chicken.
3. Cut through the bones and remove the back bones entirely.
4. Splay open the chicken, so the chicken is flat and the skin is facing up.
5. Lay the chicken in a large, shallow, ovenproof dish or frying pan, skin side up.
6. Mix the rub ingredients and spread all over the chicken skin.
7. Cook the chicken for 20 minutes uncovered.
8. Continue to cook the chicken for about 20–25 minutes more until the internal temperature reaches 165°F and the juices run clear.

Rule of thumb: What is a spatchcock chicken, you ask? The back of the chicken is removed and ideally, the thigh bones are also removed. A whole spatchcocked chicken can be laid flat across a pan or baking sheet and cooked quickly and evenly. The denser a piece of meat the longer it will take to cook it.

With spatchcocking, the seasoning reaches all parts of the chicken, and every bite will be delicious.

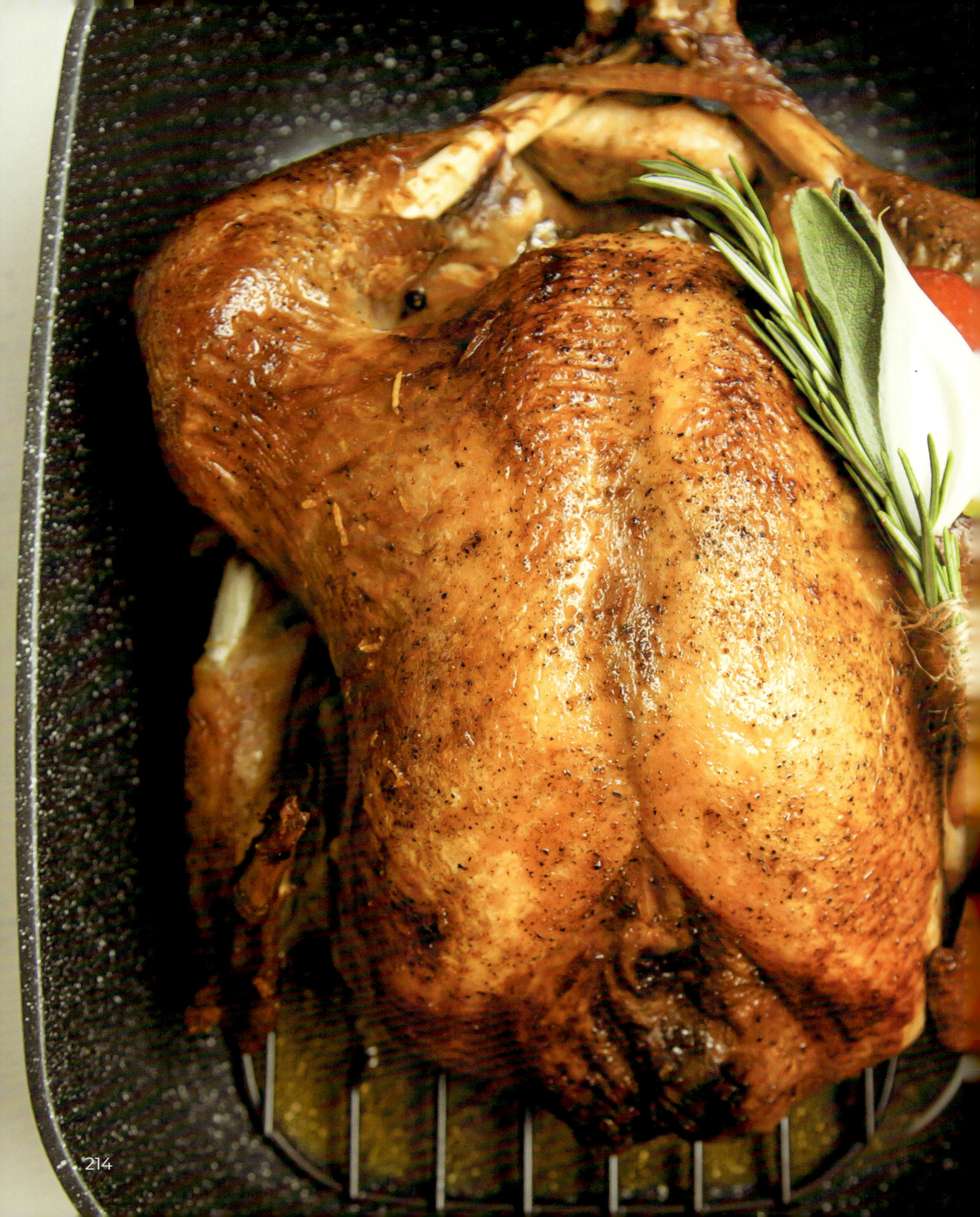

MAPLE SAGE ANYTIME TURKEY

A rub sprinkled on the turkey a few days ahead means no more work until roasting. Slip the turkey straight into the oven and finish it with glaze.

1 14-lb frozen turkey
3 tbsp coarse kosher salt
3 tbsp maple sugar or packed brown sugar
1 tsp dried, crushed, or ground sage
½ tsp freshly ground black pepper
2 sprigs rosemary
2 sprigs sage
1 medium onion, peeled and cut into quarters

1 medium apple, cored and cut into quarters
2 cups hot water
½ cup maple or table syrup
¼ cup unsalted butter
4 tsp finely grated orange peel
1 tsp ground chipotle chili pepper
1 tsp whole black peppercorns

Partially thaw the turkey, at least 1 to 2 days in the fridge. Most of the skin should no longer be icy and should give to the touch.

Remove the packaging. Remove the neck and giblets from inside of the turkey. Drain and pat the turkey dry. Loop kitchen string around the drumsticks and tie them securely to the tail.

PREPARE THE TURKEY

1. Make the rub. In a small bowl, mix the salt, sugar, sage, and black pepper. Rub this mixture evenly over the turkey.
2. Put the rosemary, sage, onions, and apples into the turkey's cavity.
3. Ball up some aluminum foil and push it into the cavity to keep the moisture from escaping.
4. Place the turkey on a rack in a roasting pan. Cover loosely with plastic wrap. Refrigerate 1 day more or until fully thawed. Remove the plastic wrap the night before cooking to let the air dry out the turkey's surface slightly before cooking.

COOKING THE TURKEY

1. Preheat the oven to 450°F.
2. Transfer the turkey to the oven. Reduce the temperature to 400°F. Roast uncovered, 1 hour.
3. Reduce the oven temperature to 350°F and cook for another hour. If the turkey breast browns too quickly, cover it loosely with foil.
4. Make the glaze. In a small saucepan, heat the maple syrup, butter, orange peel, chipotle, and peppercorns until warm. Whisk to combine.
5. For the next 2 hours of roasting, brush the turkey with the glaze every 20 minutes until a meat thermometer inserted into thigh (not touching a bone) registers 175°F.
6. Remove the turkey from the oven. Tent loosely with foil and let the turkey stand 15 minutes.
7. Reserve the turkey drippings and reserve any leftover glaze. Skim the fat from both the dripping and the glaze and use to make the gravy (see p. 89).

Rule of thumb: Brown sugar and syrup combined with the protein in the turkey skin undergo a Maillard reaction to develop the golden-brown colour.

Tenting the fully cooked bird (or any meat) is a good idea before cutting into its juiciness. The heat continues to increase the internal temperature of the turkey for a few minutes even after you remove it from the oven. But that heat increase quickly changes when the bird is exposed to colder ambient temperatures and the heat begins to move outward, along with hot moisture. Tenting with foil or a cover creates a barrier of warm air between the foil and air, so less and less heat and moisture escapes in that fast exodus caused by the large temperature differential—the difference of bird temperature to air temperature. Tenting slows the moisture loss and temperature drop. So have some foil handy when you bird reaches its final temperature. Bye-bye, dry birdie!

For more on Maple Sage Turkey, check out the "How To" video on YouTube.

TURKEY TETRAZZINI

MAKES 4–6 SERVINGS

You can substitute chicken for turkey, but I love this dish right after Christmas for any potluck. It tastes wonderful with all the flavours and dry white wine to complement the turkey, which doesn't dry out, even if this dish is rewarmed. It's a versatile recipe where there is plenty of room to freestyle the type of pasta, sauce, and cheese. Have fun!

2 cups cooked turkey meat, 1" diced
2 tbsp butter or margarine
1 cup chopped onion
1 cup chopped celery
1 cup chopped mushrooms
2 cups béchamel sauce (see p. 84)
½ cup dry white wine
2 tbsp Worcestershire sauce
2 cups cheddar cheese, grated
½ lb dried spaghetti (enough for 4–6 people)
Salt and pepper to taste

1. Preheat the oven to 350°F.
2. Prepare a greased ovenproof glass casserole dish.
3. Melt the butter in frying pan. Brown the turkey separately and remove from the pan. Then use the same pan to sauté the onions, celery, and mushrooms.
4. Boil water and cook the spaghetti to al dente. Drain.
5. Add the noodles to the bottom of the casserole dish.
6. Put the sautéed onions, celery, and mushrooms on top of the noodles.
7. Place the browned turkey on top of the vegetables.
8. Mix the béchamel sauce, milk or wine, and Worcestershire sauce together. Pour over the turkey.
9. Sprinkle the grated cheese on top.
10. Bake, uncovered for 35–45 minutes until golden brown and bubbling.

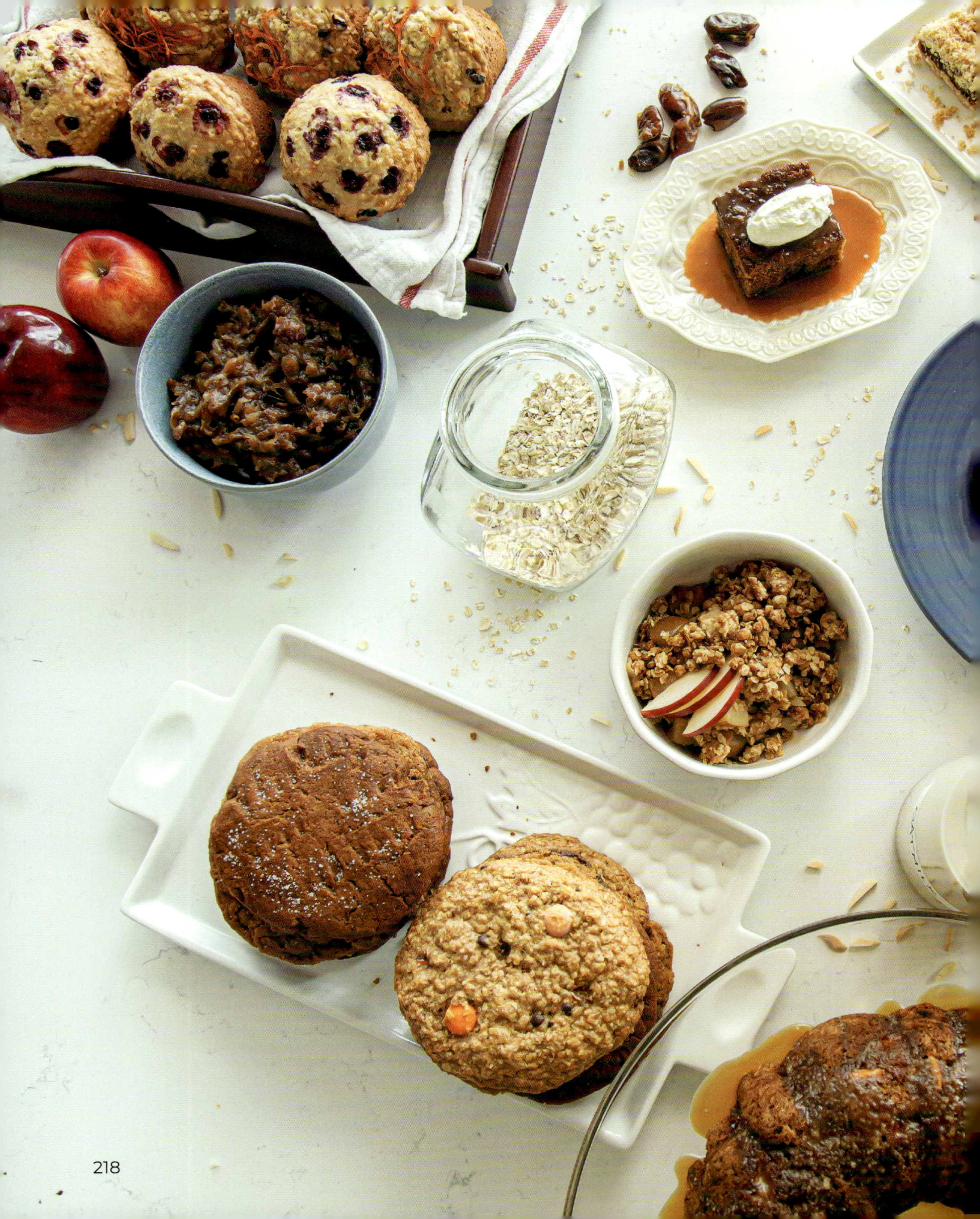

12. Nutritious and Sweet

MY STORY: SHARING

When my daughters were in grade school, I felt it was my responsibility to ensure they had the best nutrition possible. I too love to have something sweet that is also a healthy item, but I don't want to eat them all. For me, what my girls ate was easy and all in my control—or so I thought. The girls would take a couple of homemade berry oatmeal muffins packed with flavour and healthfulness freshly made from the oven in the morning for a school snack. Once there, they would attempt to trade the muffins for commercial cookies and icing dip. Employing mise en place, I can make muffins and date squares from scratch in 25 minutes, the time it takes to hydrate oatmeal and mix it together, pop it in the oven, run up the stairs, get a shower, get dressed, and then pull the muffins hot out of the oven. Later, these muffins were more welcomed by my colleagues in food manufacturing facilities and management meetings. There were complaints at most locations when I arrived without my famous date squares or healthful muffins, so I generally brought a pan to share.

THE MAGIC OF SCIENCE: FRUIT AND OATMEAL

Oatmeal and dried fruit, such as dates, apples, and berries, offer wonderful versatility and functionality when added as an ingredient to many breakfasts, desserts, and baked goods. These fruits have components that, when added to liquid, continue to absorb that liquid over time. When you apply heat and mix, the fruits absorb liquid more quickly. Oatmeal and dried fruits are both humectants meaning they will take on moisture from around them. These high-fibre ingredients substitute for water-holding ingredients such as flour and starch in many foods, holding moisture and creating texture and thickening for soups, sauces, and dressings.

Pectin, the soluble fibre in the peel and pulp of various fruit, creates the magic in thickening jams and jellies. Pectin is a hydrocolloid whose complex structure unravels with heat in the presence of acid, sugar, and water and creates viscosity. The result is a smooth gel texture when cooled. Of course, the firmness of the gel is dependent on the ratios of the pectin from the fruit, acidity, level of sugar, moisture, and hardness of the water. Calcium in hard water can have a positive effect on the strength of the gel. Different fruits offer different levels of pectin. Apples, berries, and dates, for me, are the most versatile, most functional, and most nutritious, and they're used widely in many cuisines. Fresh fruit and dried fruit have natural sugars as well as pectin and inherent moisture that allows them to impart their own sweetness and distinct flavour. Manufacturers of pectin extract it from the pulp and peel of citrus fruits such as lemons and oranges and from apple peels. For decades, oatmeal's positive health effects in lowering blood cholesterol and glucose levels and in cancer prevention have been associated with Beta glucan, a soluble fibre. Oatmeal consists mainly of starch, fibre, and protein. When water is added to oats, the fibre naturally absorbs some of the moisture through swelling. With heat, the starch molecules unwind and hold the water molecules more tightly, and the texture changes to a soft, moist consistency. With agitation and prolonged heating, the soluble fibre fraction takes on more moisture and, together with the fully hydrated starch, creates a gluey smooth texture and translucent appearance, particularly when it is cooled. This thickening property of the fibre and starch components of oats means they can hold three times their weight in water when cooked and cooled, giving softness and elastic spring to any cake or dessert and preventing them from drying out or going stale quickly.

- Cooking Dates (and Other Fruit)
- Traditional Date Squares
- Sticky Toffee Pudding and Toffee Sauce
- Apfelkuchen Apple Cake
- Cranberry Compote
- Partridgeberries and Apple Jam
- Basic Breakfast Oatmeal
- Oat Milk
- Soft Oatmeal Muffins
  - Muffin Freestyling
- Fruit Crisp
- Cookies
  - Basic Cookie Recipe
  - Cookie Freestyling
  - Polka Dot Cookies

A party without cake is just a meeting.
-Julia Child

COOKING DATES (AND OTHER FRUIT)

MAKE 2 ⅓ CUPS

1 cup dates, pitted and chopped
⅔ cup water

STOVETOP METHOD

1. In a saucepan, combine the dates and water. Bring to a boil and simmer until thickened and soft.
2. To use as is: Strain and cool the mixture before using it in recipes.
3. Use as a paste: Stir vigorously until the mixture becomes a thick paste.
4. Use later: Refrigerate the mixture for up to 1 week or freeze for 2 months.

MICROWAVE METHOD

1. Put water and dates in a microwaveable bowl. Cook on high for 1 minute.
2. Stir the date mixture vigorously with a wooden spoon. Cook in the microwave on high for 30 seconds more.
3. Beat vigorously and repeat an additional 30 seconds on high, then beat again until the mixture is a paste. Do not overcook all at once as the dates will burn. They need to continually absorb the water.

Rule of thumb: Dates absorb the excess moisture and take on more moisture due to their inherent characteristics. Dates are today's alternative to sugar, so if dates are pureed, each bite has natural sweetness.

Other fruit such as dehydrated apples can be rehydrated this way too.

HOW TO HYDRATE DRIED FRUIT FOR BATTER

Dried fruit needs to be rehydrated before you add it to batter. Otherwise, the dried fruit will steal the moisture meant to make the cake or muffins moist. Soak ⅔ cups dried cranberries, cherries, or raisins in 1 cup boiling water for 15 minutes. Strain well.

TRADITIONAL DATE SQUARES

MAKES 16 SQUARES

These squares are my signature go-to for any occasion. They bring up a little nostalgia and comfort for many. They are sweet enough to curb that sweet-tooth craving and healthy enough being low in sugar and high in fibre. You can have more than one square if you like or have them for breakfast with a little yogurt.

¾ cup brown sugar, lightly packed
1 ⅓ cup flour
½ tsp baking soda
2 cups rolled oats

⅔ cup butter or margarine, melted
2 ⅓ cup cooked dates (see p. 221 for instructions on how to cook dates)

1. Preheat the oven to 325°F.
2. Lightly grease an 8"x 8" baking pan.
3. Combine the flour, sugar, baking soda, and oats in a mixing bowl.
4. Melt the margarine and add to the flour–oat mixture. Mix well until crumbly.
5. Press half of the crumbs into the prepared pan.
6. Spoon the cooked dates on top of the pressed crumb layer one teaspoon at a time, distributing the date mixture evenly and smoothly across the pan.
7. Cover with the remaining crumbs and pat them smooth.
8. Bake for 35–40 minutes or until golden brown.
9. Cool and cut into bars.
10. These date squares hold for a few weeks in an airtight container. They also freeze well.

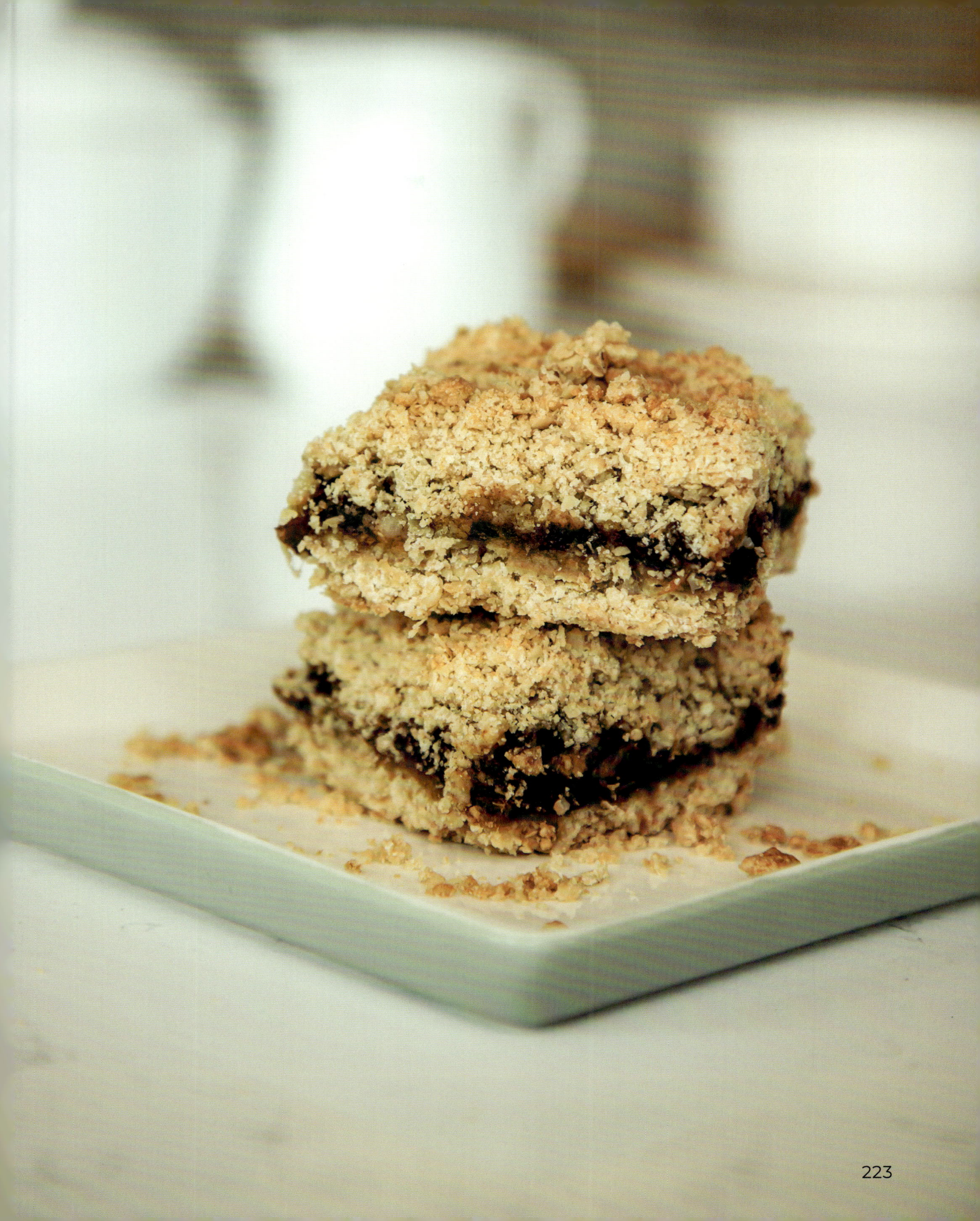

STICKY TOFFEE PUDDING

MAKES 6 SERVINGS

This is another Java Jack's signature dish that has been delighting guests for years now. It's so popular we have packaged it up in a bakery mix for you to make at home. The dates and dry ingredients are included, and all you need to do is add all the wet ingredients you likely have in your pantry. You can make this ahead of time and freeze it.

1 cup flour
1 tsp baking powder
1 tsp baking soda
¼ cup unsalted butter or margarine, softened

¾ cups granulated sugar
1 large egg, lightly beaten
1 tsp vanilla
1 ½ cups dates (uncooked)

1. Preheat the oven to 350°F.
2. Lightly grease an 8" square baking pan.
3. Chop the dates and place them in a bowl.
4. Add boiling water and baking soda to the dates, then stir until dates are soft and water is almost absorbed, about 5 minutes. Let stand for 15 minutes more to fully hydrate the dates. (This can be done a day ahead.)
5. In a mixer, cream the butter and sugar until fluffy.
6. Add the egg and vanilla. Beat until smooth.
7. Gradually beat in the flour and baking powder.
8. Add the date mixture and mix with a spatula.
9. Bake until the pudding is set and firm, about 30–35 minutes.
10. Let cool and cut into nine squares. Top with lots of toffee sauce and whipped cream.

Rule of thumb: Cut the cooled pudding with a hot, damp knife and clean the knife in between cuts. The soft moist date cake sticks to the knife otherwise.

Make the pudding and freeze the whole pan's worth. It is easier to cut the pudding when frozen. Reheat the pudding in the microwave on high for 40 seconds per piece, then top with toffee sauce.

TOFFEE SAUCE

MAKES 2 CUPS

½ cup heavy cream (35% fat whipping cream)

1 cup light brown sugar
½ cup unsalted butter

1. Add everything to a saucepan. Stirring constantly, heat on medium until the sticky saucy mixture begins to bubble.
2. Keep stirring and let the sauce thicken for 3–5 minutes until it adheres to the back of the spoon. It's ready!

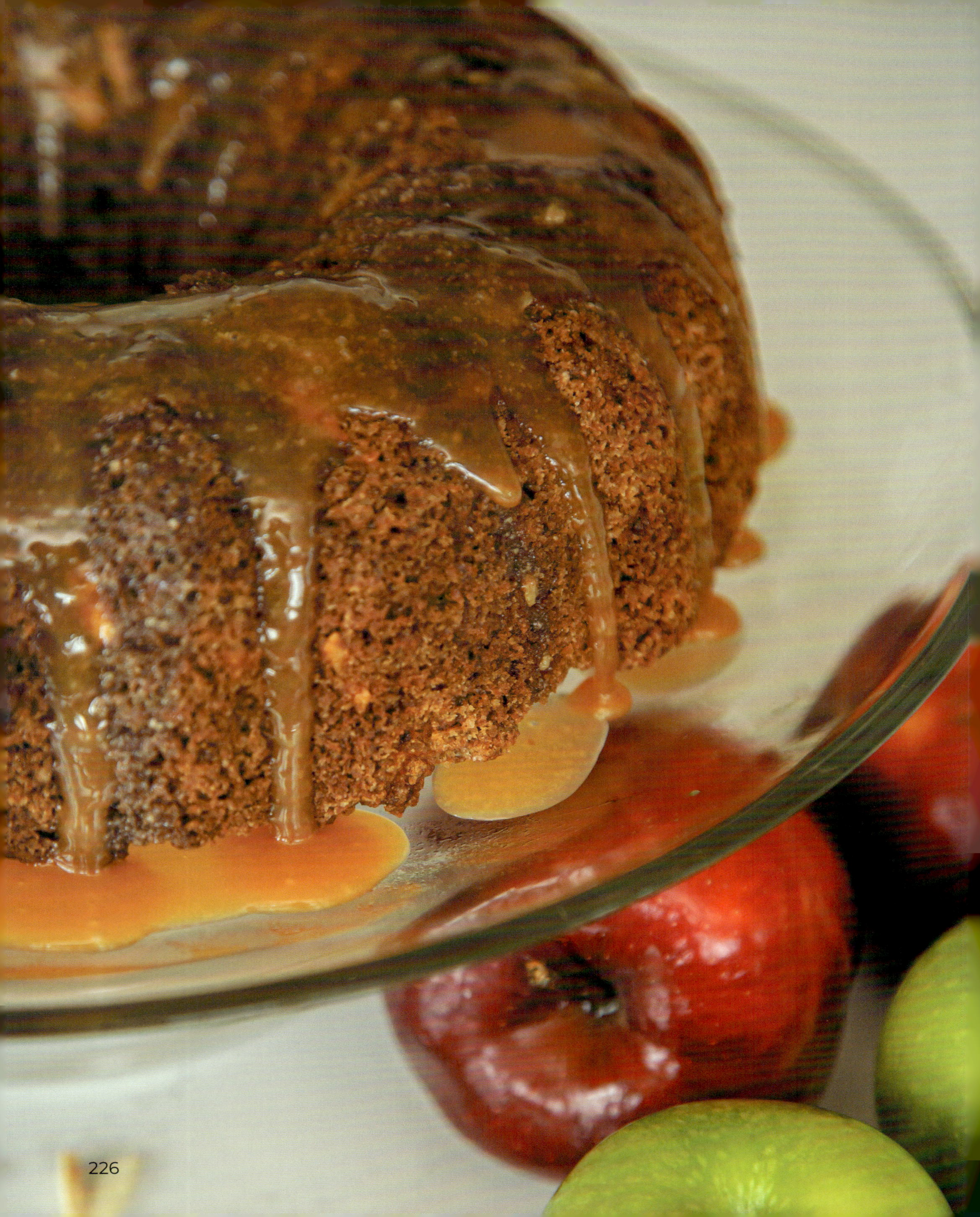

APFELKUCHEN APPLE CAKE

MAKES ONE BUNDT CAKE

On one of my business trips to Germany, I discovered Apfelkuchen. I tried it every day and everywhere they offered it. It is as varied a dessert there as are fish cakes here. I have a few recipes, and one that is more like an apple pie, which my father loves. That recipe uses his crop of backyard apples. This recipe is a decadent moist coffee cake I have made with both fresh apples and dehydrated ones. The almond and vanilla flavourings and the crunch from the slivered almonds makes it more special. Drizzle the cake with toffee sauce (see p. 224), and it is divine.

3 cups flour
2 tsp baking soda
¾ cup brown sugar
¾ cup sugar
1 ½ tsp salt
3 tsp cinnamon
1 ½ cups light vegetable oil
2 eggs

1 tbsp vanilla
½ tsp almond extract
½ cup toasted slivered almonds
3 cups apples, peeled, cored, diced into 1" chunks (or 1 cup dehydrated apples, chopped into ¼" pieces and soaked in 1 cup of boiling water for 30 minutes.)

1. Preheat the oven to 350°F.
2. Grease a Bundt pan by spraying generously with cooking oil.
3. Line the bottom of the pan with parchment paper.
4. Whisk the egg and vegetable oil together. Add the vanilla and almond extract.
5. In another bowl, mix the flour, sugar, brown sugar, cinnamon, baking soda, and salt together.
6. Combine egg and flour mixtures and stir until just incorporated.
7. Add the apples. Mix well.
8. Fold in the almonds.
9. Pour the batter into the Bundt pan, spreading evenly.
10. Bake for 50 minutes to 1 hour at 350°F.
11. Cool for 10 minutes, then remove from the pan and place on a platter.
12. Pour 1 cup of warm toffee sauce (see p. 224) over the cake while it is warm.

CRANBERRY COMPOTE

MAKES 2 CUPS

Make this sauce ahead of time for your Thanksgiving turkey or use it on your toast in the morning. My daughter freestyled this recipe to accompany our Christmas dinner and added a teaspoon of finely grated orange peel. Making this compote is so easy you will never bother to buy the canned version again.

2 cups cranberries
(or partridgeberries), fresh or frozen

1 cup sugar
1 tbsp cinnamon

1. Place all the ingredients in a saucepan.
2. Put the heat on low and let the mixture heat, stirring occasionally.
3. When the mixture starts to bubble and thicken so it adheres to the back of a spoon, remove the saucepan from the heat.
4. Let the mixture stand for 5 minutes before putting into jars.

PARTRIDGEBERRIES AND APPLE JAM

MAKES 2 CUPS

Partridgeberries are internationally known as lingonberries and grow in the dry acidic soil on the barrens in Newfoundland and Labrador. These berries are a relative to cranberries and have similar thickening and gelling properties when boiled down to a jam

2 cups partridgeberries, fresh or frozen
½ cup sugar

1 tbsp cinnamon
1 cup chopped or grated apples, peel on

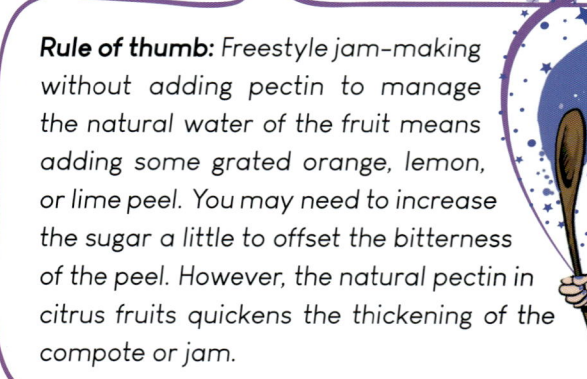

1. Place all the ingredients in a saucepan.
2. Put the heat on low and let the mixture heat, stirring occasionally.
3. When the mixture starts to bubble and thicken so it adheres to the back of a spoon, remove the saucepan from the heat.
4. Let the mixture stand for 5 minutes before putting into jars.

Rule of thumb: Freestyle jam-making without adding pectin to manage the natural water of the fruit means adding some grated orange, lemon, or lime peel. You may need to increase the sugar a little to offset the bitterness of the peel. However, the natural pectin in citrus fruits quickens the thickening of the compote or jam.

BASIC BREAKFAST OATMEAL

MAKES 1 SERVING

You can see the magic of oatmeal's thickening and water-holding properties in front of your eyes while making your breakfast. My father has his oatmeal with milk and blueberries every morning.

½ cup oats
1 cup milk

STOVETOP
1. Place the oats and milk in a small saucepan. Mix well, stirring constantly. Cook on medium-high heat for 10–15 minutes until the mixture starts to thicken and large bubbles form.
2. Remove and serve.
3. Add some blueberries, sliced almonds, cranberries, dates, or whatever you like

MICROWAVE
1. Mix the ingredients in a bowl. Cook on high for 45 seconds, then stir for 20 seconds.
2. Return the mixture to the microwave and cook on high for another 20–30 seconds. Stir and serve.
3. Add some blueberries, sliced almonds, cranberries, dates, or whatever you like.

OAT MILK

MAKES 8 CUPS

1 cup rolled oats, oat flour, or steel-cut oats
8 cups water
2 tbsp sunflower oil (optional)

1. Soak the oats in water at room temperature for an hour.
2. Add the oats and water to a blender.
3. Blend on high for 4 minutes, pulsing every minute.
4. Pass the mixture through a very fine mesh strainer to extract most of the oat bran.
5. Transfer the liquid to a saucepan and heat to 60°C for 4 minutes until the oat solution starts to thicken.
6. Chill immediately and refrigerate for up to a week.

For additional body and silkiness, add the optional 2 tbsp of a light-bodied oil, such as sunflower oil, and blend for 1 minute in the blender or with an immersion blender. Then chill.

Rule of thumb: If the oat milk becomes too thick and gluey, add 6 slices of a ripe banana or a few cubes of ripe mango during the heating stage, and let their active enzyme (amylase) do its work for 30 minutes. Remove the fruit before chilling. The milk will be sweet and the glueyness will go away.

SOFT OATMEAL MUFFINS

MAKES 12 MUFFINS

1 cup oats
1 ¼ cup milk
1 egg
¼ cup melted butter
1 tsp vanilla
1 cup flour
¾ cup brown sugar
1 tsp baking powder
½ tsp baking soda
½ tsp salt
½ tsp cinnamon
Fruit, nuts, chocolate chips
(see Muffin Freestyling)

1. Preheat the oven to 400°F. Grease a muffin pan with spray oil or place muffin liners in each muffin cup.
2. Mix the oats in the milk vigorously, then let stand for 15 minutes.
3. Beat the egg and add the melted butter.
4. Add to the eggs and butter to the oats and milk, and stir until well mixed.
5. Combine the rest of the ingredients together in a large bowl.
6. Add the oat mixture to the dry ingredients. Stir just until moistened.
7. Fold in any inclusions, such as chocolate chips, berries, nuts, or other fruit gently.
8. Portion equally into the muffin cups.
9. Bake for 20–22 minutes and serve warm. Reheat in the microwave on high for 20 seconds.

Rule of thumb: *The soft moistness of these muffins is in the magic of the oats.. Let the oats sit for at least 15 minutes so the oats absorb the milk, the starches swell, and the oats are not competing with the flour for liquid during the cooking process.*

For more on Oatmeal Muffins, check out the "How To" video on YouTube.

MUFFIN FREESTYLING

Chocolate Chip Add 1 cup chocolate chips.	**Berry** Add 1 cup berries, fresh or frozen.	**Jam** Portion out the batter into the muffin pan. Add a teaspoon of jam to the centre of each muffin's batter, and let the jam sink into the batter before baking.
Apple Cinnamon Add 2 apples, grated or finely chopped, 1 tbsp cinnamon, and ¼ cup brown sugar.	**Carrot Walnut** Add ⅔ cup grated carrots, an extra 1 tsp cinnamon, ¼ cup chopped toasted walnuts, and ¼ cup raisins.	**Pear Pecan** Add 2 pears, grated or finely chopped, and ½ cup chopped toasted pecans.
Apricot Poppy Seed Add ½ cup finely chopped dried apricots and ½ cup poppy seeds.	**Blueberry and Brown Sugar** Add ⅔ cup blueberries and 2 tbsp brown sugar to put on top of each muffin before baking.	**Banana and Chocolate Chip** Add 2 mashed bananas, ⅔ cup chocolate chips, and ¼ cup finely chopped walnuts.
Lemon Poppy Seed Add the zest from 1 lemon, 1 tbsp lemon juice, and ½ cup poppy seeds.	**Berry Granola** Add ½ cup blueberries and ½ cup granola.	**Cranberry Orange** Add the zest from 1 orange, 1 tbsp orange juice, and ⅔ cup cranberries.
Coconut Lime Substitute the cinnamon with 1 tsp coconut extract, ⅔ cup coconut, zest from a lime, and 1 tbsp lime juice.		**Peanut Apricot** Add ½ cup finely chopped dried apricots and ½ cup chopped peanuts

233

FRUIT CRISP

MAKES ABOUT 6 SERVINGS

This variation of a crisp is easy and delicious. Consider substituting any berries or fruit or just make it all apple or all pear. The recipe looks like it asks for a lot of fruit, but the bulkiness of the uncooked fruit reduces as it cooks.

6 cups pears, unpeeled, cored and chopped
2 cups apples, unpeeled, cored and chopped
2 tbsp flour

2 tbsp brown sugar
2 tbsp lemon juice
1 tsp cinnamon

TOPPING

1 ½ cup quick oats
½ cup brown sugar
⅓ cup flour

Pinch nutmeg
⅓ cup melted butter

1. Preheat the oven to 350°F.
2. Combine the fruit with the lemon juice to prevent the fruit from browning, then toss with the other dry ingredients.
3. Place the mixture in the bottom of greased, individual oven-safe dishes (or one medium 3-quart baking dish).
4. Combine the oats, brown sugar, flour, and nutmeg together for the topping. Add the melted butter and mix.
5. Distribute the topping evenly over the fruit. Bake for about 1 hour until golden.

Rule of thumb: The juices from the fruit on the bottom moisten and flavour the oatmeal in the topping.

Sometimes, depending on the choice of apple and ripeness of pear, the pectin in the peel may not be sufficient to hold all the inherent fruit juices in the bottom, so a little flour may help hold the juices to create that finished smooth soft texture, hot or cold.

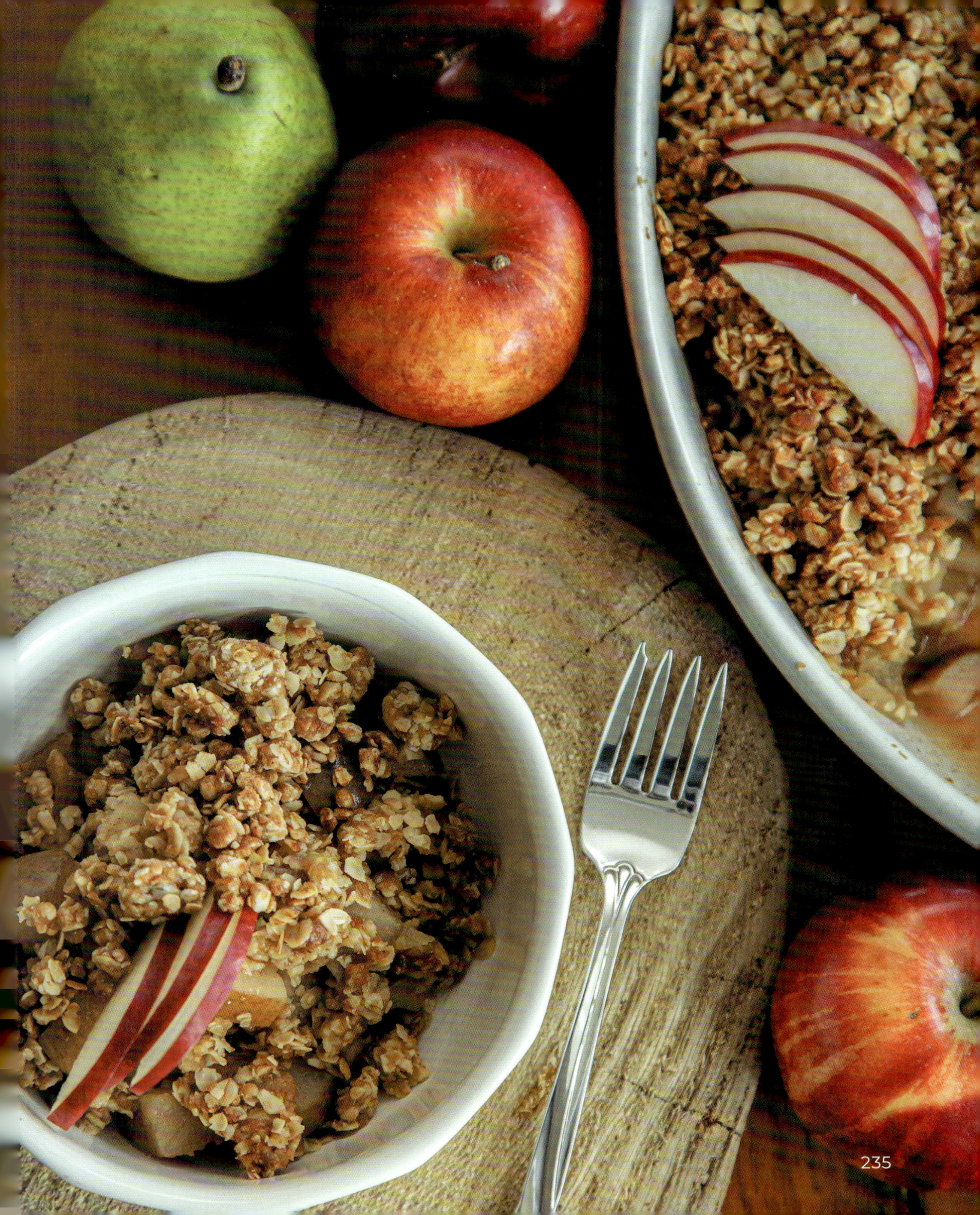

COOKIES

I love cookies! I had to add this. A good homemade cookie recipe is hard to find. Here is a foolproof base recipe from which all others can be derived. Check out the freestyling options with raisins and oatmeal or freestyle your own.

BASIC COOKIE RECIPE

MAKES 12 COOKIES

½ cup unsalted butter, softened
½ cup sugar
½ cup brown sugar
1 egg

1 tsp vanilla
1 ⅓ cup flour
½ tsp baking soda
½ tsp salt

1. Preheat the oven to 350°F.
2. Prepare baking sheets with parchment paper.
3. In a mixer, cream the butter, sugar, and brown sugar so it is well distributed and smooth.
4. Add the egg and vanilla, and beat well until fluffy.
5. Mix the flour, soda, and salt together in a separate bowl.
6. Add the flour mixture to the butter, sugar and egg mixture.
7. Mix everything together until it is uniform. Scrape the bottom of the bowl to get any dry ingredients that were missed and mix in.
8. Add inclusions such as nuts or chocolate chips, folding in for even distribution.
9. Roll 2 tbsp of dough into a ball. Place on the baking sheet and press each cookie slightly so they are uniform. The dough will not rise a lot, and it will spread, so place the cookies 2" apart.
10. Bake for 10–12 minutes until just brown on the bottom and soft in the centre. Bake longer for a crispy cookie.

Rule of thumb: Don't have a mixer or don't want to haul yours out? No worries. Mix the softened butter and sugars together with a wooden spoon until well mixed and smooth.

In another bowl, whisk the egg until frothy and the colour is light yellow, then add the vanilla. Whisk a bit more, then add the eggs to the butter-sugar mixture. Mix well until it is all combined, then mix in the dry ingredients to create the cookie dough.

COOKIE FREESTYLING

Chocolate Chip
Add ¾ cup chocolate chips.

Peanut Butter
Add ½ cup peanut butter, ¼ tsp cinnamon, and an extra ½ tsp baking soda.

Snickerdoodle
Add extra ⅓ cup flour, 1 tsp cinnamon, and ½ tsp baking powder.

Oatmeal Raisin
Replace ⅔ cup of flour with oats, and add ½ cup plumped raisins (see p. 221).

Molasses Ginger
Replace the sugar with molasses, and add 1 tsp ground ginger. Roll in white sugar before baking.

Monster
Add ½ cup peanut butter, replace ⅔ cup flour with oatmeal, and add ½ cup mini chocolate chips and ½ cup Smarties.

POLKA-DOT COOKIES

MAKES ABOUT 2 DOZEN

¾ cup sugar
¾ cup butter or margarine
1 egg
1 tsp vanilla
½ tsp salt
1 tsp mace
1 cup flour
1 cup quick oats
¾ cup toasted walnuts, finely chopped
1 cup chocolate chips

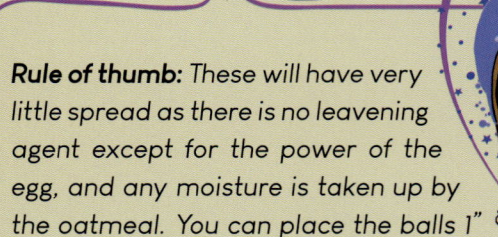

Rule of thumb: These will have very little spread as there is no leavening agent except for the power of the egg, and any moisture is taken up by the oatmeal. You can place the balls 1" apart on the baking sheet.

1. Preheat the oven to 375°F.
2. Line a baking sheet with parchment paper.
3. In a mixer, cream the butter and sugar.
4. Add the egg and vanilla. Mix until fluffy.
5. In another bowl, mix the flour, oatmeal, salt, and mace together.
6. Combine the dry ingredients and the sugar-butter-egg mixture together and mix until everything is equally distributed and a dough forms.
7. Fold in the nuts. Let stand 10 minutes for the oats to swell and hydrate slightly.
8. Shape the dough into 1" balls, Place the balls on the parchment paper and press each down in the centre with a spoon or your finger.
9. Bake for 5 minutes and then remove from the oven.
10. Press down in the centre of each cookie again. Return to the oven and bake for another 5–7 minutes.
11. Melt the chocolate chips in the microwave on high for 30 seconds. Stir and microwave again for another 30 seconds or until the chocolate is smooth.
12. Filll the centres of the cookie with ¼ tsp of melted chocolate.
13. Let cool.

APPENDIX A

THERMOMETER CALIBRATION PROCEDURE

ICE SLURRY METHOD
1. Fill a large drinking glass with crushed ice.
2. Fill with cold water just below the ice.
3. Leave for 2 minutes and stir for 10 seconds. This is an ice slurry.
4. Place the probe of the thermometer to be tested into the ice slurry and slowly stir to allow the temperature reading on the display to stabilize.
5. The temperature displayed should be 0°C (32°F) +/- 1°C.
6. If the temperature is within +/-1°C (2°F), the thermometer meets the requirements of the standard. Proceed to use it.
7. If the temperature displays greater than +/- 1°C (2°F), it should be replaced (or recalibrated professionally).

GLOSSARY

Acidulant: a food substance used to preserve food by increasing acidity
Aeration: action of introducing air into ingredients to make them fluffy and light
Agitation: action of mixing, shaking, tumbling, or blending things together
Aquafaba: liquid from soaking chickpeas or the juice from canned chickpeas
Aqueous: water, water-like, made from water
Au jus: juices or drippings, as is, from the cooking of meat

Bind: to attach, connect, join, or combine two or more substances
Body: having some viscosity
Bouquet garni: French for a bundle of herbs wrapped in cheesecloth and tied with string, mainly used to prepare stocks and soups, replace and with that is removed after cooking.
Bundt: shape of cake pan that allows more of the cake to be in contact with the edges of the pan, and therefore the heat is transferred more evenly
Butterfly: to dissect into two pieces that are held together in the centre

Calibrate: to assess, set, or adjust readings to accurately match the standard
Carcass: remains of a cooked bird or other animal after all the edible parts have been removed
Cellulose: main substance in the walls of plant cells, helping plants to remain stiff and upright
Coagulate: to change something from a fluid to a solid state
Cruciferous: hearty vegetables such as cabbage, cauliflower, Brussel sprouts, kale, broccoli from plants in the family Cruciferae

Deglaze: cooking process using liquid to remove browned food from the bottom of the pan to flavour sauce
Dehydrate: to remove moisture or the process by which moisture is removed
Diffusion: movement of molecules of a substance or gas through a semipermeable barrier from an area of higher concentration to an area of lower concentration
Dissolve: when something melts, turns into, or becomes part of a liquid form
Dot with butter: butter that is cut in small cubes, distributed evenly over the surface
Dredging: coating moist foods with a dry ingredient such as flour
Dressed: the body of an animal that has been cleaned to remove the entrails and fur (carcass)
Dutch oven: a large pot with a cover, typically with two handles that can be used on the stove and in the oven

Elasticity: stretchiness
Enzymes: proteins that help to regulate the rate of chemical reactions without being altered themselves.
Equilibration: condition where two opposite things are in balance
Extract: to pull out or withdraw something by physical or chemical process

Flambé: technique to cover food with liquor and set it afire and imparts a subtle liquor flavour that compliments desserts and sauces.
Fold into: stir gently with a spatula to incorporate an ingredient so to maintain aeration and lightness
Fundamentals: basic or most important rules, ideas, or truths that guide actions and thinking

Granule: small compact particle
Glycogen: a stored form of glucose (i.e. sugar) in liver and muscles for a quick boost of energy

Humectants: ingredients that promote moisture retention (water holding)
Humidity: the amount of moisture in the atmosphere
Hydration: the process used to help material absorb water

Intermuscular: situated between and separating muscles
Intramuscular: situated within the muscle
Italian seasoning: a blend of herbs consisting of basil, oregano, thyme, marjoram and rosemary

Kinetic energy: energy existing by being in motion

Leavening: ingredient to make a dough, cookie, or batter rise
Lecithin: natural emulsifier that is in egg yolks, mustard, and other plant and animal foods.

Maillard reaction: browning that occurs in foods when proteins chemically react with simple sugars
Mechanical action: movement of mechanical parts that impact or shear
Molecular: group of atoms that is the smallest unit of a chemical compound
Myofibrillar proteins: types of protein that binds water and fat together in a meat product

OMG: expression of excitement and amazement; stands for "oh my gosh"
Opaque: cloudy such that light does not shine through
Optimize: arrange or use something in a way that improves efficiency or effectiveness
Oxidation: the result of adding oxygen through a chemical process
Oxygen transmission rate: how quickly oxygen gas permeates steadily through film

Percent of solids: the amount of solids, dissolved or undissolved in a liquid
Persisting bacteria: bacteria that survive exposure to disinfectant, antibiotics and other stresses
pH: a scale from 1 to 14 that indicates something's of acidity, with 1 being very acidic and 14 being very caustic or basic. The pH of water is 7, a neutral pH
Poach: to cook (an egg) without its shell in boiling water or to cook by simmering in a small amount of liquid
Post-mortem: after death
Preservative: substance or chemical that helps keep food from deteriorating
Pulverize: to reduce to fine particles
Purge: to physically remove or force something out; free liquid from meat

Residual: leftover or waste, extra
To the ribbon: result of whisking eggs until they are light yellow and thick and move slowly back in place in the bowl as the whisk sweeps back and forth
Rigor mortis: stiffening of the muscles of a body after death, lasting up to four days
Root vegetable: the fleshy enlarged root of a plant used as a vegetable, e.g., a carrot
Rule of thumb: using experience or practice as a way to judge a situation or guide actions

Simmer: a temperature that's just below boiling while still heating
Soluble: the ability of a substance to be dissolved in water
Spatchcock: split open to prepare for cooking, roasting, grilling
Splay: spread out flat and apart
Subcutaneous: just underneath the skin
Surface area: the total of the outside part of something

tbsp: tablespoon
Tenting: loosely covering food to keep it warm and still allow air flow
Translucent: light can shine through, clear
tsp: teaspoon
Tuber: fleshy root vegetable that grows underground and potentially able to produce a new plant

Uniform: when there is no difference between two things; they are the same

Viscosity: degree of thickness; resistance to flow

Yield: quantity or result that can be produced, grown, or delivered

REFERENCES

Barbut, S. (2015). *The science of poultry and meat processing.*

Bourlieu, C., Astruc, T., Barbe, S., Berrin, J. G., Bonnin, E., Boutrou, R., & Paës, G. (2020). Enzymes to unravel bioproducts architecture. *Biotechnology Advances, 41*, 107546.

Devillier, J. (2019). *The New Orleans Kitchen: Classic Recipes and Modern Techniques for an Unrivaled Cuisine.* Lorena Jones Books.

Exploratorium. (2018). Science of cooking. San Francisco, CA.

Fennema, O. R. (1996). Water and ice. *FOOD SCIENCE AND TECHNOLOGY-NEW YORK-MARCEL DEKKER-*, 17-94.

Forrest, John C., Aberle, E.D., Hedrick, H.B., Judge, M.D., Merkel, R.A. (1975). *The Principles of Meat Science.* W.H. Freeman and Company, New York.

Hultin, H. O. (1984). Postmortem biochemistry of meat and fish.

Olson, A. Olson, M. (2011). Anna and Michael Olson Cook at Home: Recipes for Everyday and Every Occasion. Whitecap Books Ltd.

Ricardo Cuisine. (2022). Food Chemistry. Ricardo Cuisine.

Rombauer, I. S., & Becker, M. R. (1974). *Joy of Cooking: The All-Purpose Cookbook.* Bobbs Merrill Company, Inc.

Sayar, S., Jannink, J. L., & White, P. J. (2007). Digestion residues of typical and high-beta-glucan oat flours provide substrates for in vitro fermentation. *Journal of agricultural and food chemistry, 55*(13), 5306–5311. https://doi.org/10.1021/jf070240z

Sun, X. D., & Holley, R. A. (2011). Factors influencing gel formation by myofibrillar proteins in muscle foods. *Comprehensive reviews in food science and food safety, 10*(1), 33-51.

Willan, A. (1989). *LaVarenne Practique.* MacMillan of Canada.

INDEX

A

aioli 45, 47, 48, 104
albumen (see egg, whites) 124, 125
almond 52, 118, 121, 144, 195, 227
almonds,
  how to toast 118
amylopectin 64, 99
amylose 64, 99
apples,
  apple pear crisp 234
  cake, Apfelkuchen (also see cake) 227
  jam 228
  muffins 232
  partridgeberry apple 228
  slices 123, 192
asparagus 97, 98, 114, 118, 195
avocado 45, 49, 57

B

bacon 50, 53, 57, 61, 92, 118, 131, 167, 190
baking powder 7, 19, 20, 21, 22, 24, 26, 27,
    29, 32, 140, 206, 224, 225, 231, 237
baking soda 7, 19, 20, 21, 22, 29, 32,
    222, 224, 225, 227, 231, 236, 237
ballotine 198, 202
banana 20, 31, 230, 232
barley (see stew) 64, 185
batter, Orly beer 165
beans (see salads) 8, 58, 59
  also 107
béchamel (see sauces) 63, 82, 84,
    86, 87, 88, 90, 111, 208, 217
beef,
  cooking 7, 187
  prime rib 186, 187
  seasoning 187
beets, roasting 55
benedict, lobster poached eggs 137
beta glucan 220, 245
biscuits,
  Grandmother Ivy's tea buns 22
  scones 26
blood 173, 220
blueberries 31, 229, 232

bœuf bourguignon 190
bouillabaisse 70, 75
brandy 192
bread,
  Irish soda 29, 32
  naan 204, 206
  quick (see loaves) 29
broth 61, 64, 65, 70, 71, 73, 75, 80,
    107, 153, 158, 167, 190, 194
browning, Maillard reaction 20, 127, 216, 242
burgers, super juicy 197, 198, 200
buttermilk, how to make 29

C

cabbage,
  green 108, 114, 121, 183, 241
  savoy 108
Caesar (see salad, dressing) 48, 53
cake,
  apple 227
  bundt 227, 241
calibration, thermometer 239
carcass 6, 64, 67, 173, 241
cellular 77, 97, 98
cellulose 78, 241
cheese,
  blue, gorgonzola 54, 57
  cheddar 73, 87, 88, 111, 131, 217
  cream 31, 114
  goat 55
  Gruyère 73, 86, 88
chicken,
  à la king 204, 208
  butter 204, 206
  butterfiied breast 207
  leftover 208
  noodle soup 71
  one-hour roast chicken dinner 210
  piccata 207
  spatchcock, fiat, smoked paprika
    garlic lemon 213
  stuffed chicken breast ballotine 202
chocolate,
  chips 26, 142, 231, 232, 236, 237, 238
  éclairs 138, 139

247

mousse, chocolatey 142
choux pastry 138
chowder, seafood 63, 90
chunk, chunky 71, 75, 88, 94, 95,
    100, 120, 183, 185, 227
cinnamon 26, 30, 31, 144, 227, 228, 231, 232, 234, 237
coalescing 45
cobbler, strawberry 27
cod,
  au gratin 88
  beer battered 165
  fish 63, 75, 82, 88, 90, 94, 151,
    152, 160, 161, 165, 180
  pan-fried with scrunchions 160
coleslaw 183
collagen,
  as connective tissue 172, 173, 174, 175, 179
  as gelatin, gelatinous 69
composition,
  chicken breast 173
  lean beef 173
  lean pork 173
  muscle 173
compote, cranberry 228
cookie freestyling 237
cookies,
  basic recipe 236
  chocolate chip 237
  molasses ginger 237
  monster 237
  oatmeal raisin 237
  peanut butter 237
  polka dots 238
  snickerdoodle 237
cooking,
  dates (and other fruit) 221
  fish 153
  meat 12, 179
  seafood 153
cooking methods,
  baking 179
  boiling 179
  braising 153, 180
  broiling 179, 180
  deep frying 180
  dry heat 180
  pan frying 153
  pan searing 153
  pressure cookers 180

roasting 179, 180
step 178
stewing 179, 180
water, steam 153, 154, 180
crème brûlée 123, 143
crisp,
  apple pear 234
  fruit 220, 234
croutons, how to make 53
curd, lemon 140, 141
custard 125, 138, 143

D

date squares, traditional 222
dates (see fruit) 220, 221
density, densities 10, 178
diced 100
diffusion 178, 241
dispersion 35, 37
dressing,
  creamy avocado 49
  easy Caesar salad 53
  sweet asian 42
duchesse (see potatoes) 102, 107
dumplings, for stew 24

E

egg,
  as composition 123
  as functionality 125
  gel networks 124
eggs,
  benedict, butter poached lobster 137
  frittata, spinach bacon breakfast 131
  scrambled, fluffy minute 127
eggs, how to
  boiled 129
  devilled 129, 130
  make an egg wash 189
  peeling 129
  poaching 137, 243
  temperatures, cook 7
elastin 173
emulsifiers,
  lecithin 35, 45, 125, 242
  mustard 35, 37
emulsions 35

energy,
  cold 178
  heat 177, 178, 179, 180
  kinetic 178
equilibrium,
  as equilibration of temperature 178
equipment, kitchen (see utensils) 5, 6, 7, 8, 17

F

fatty tissue, as connective matrix 175
fish,
  Arctic char 152
  cod (see cod) 151, 152
  fish cakes 161
  haddock 152
  halibut, Riesling poached 152, 156
  rainbow trout 152
  salmon, Atlantic 152
  salmon, pan-seared with chili peach glaze 155
  tilapia 152
fish, as cook temperature 7
fish cakes with million-dollar relish 161, 163
flambé 242
flour, composition 12, 64
frittata (see eggs) 125, 131
fruit,
  cobbler 27
  crisp 234
  glaze 155
  loaves 31
  muffins 232
fruit, how to hydrate
  apples 227
  dates 221
  raisins 221

G

game, wild, small (rabbit) 7, 188
garlic,
  aioli 48
  smashed potatoes 104
glaze,
  chocolate for éclairs 138
  peach chili 155
gluten 20, 82
glycogen 173, 242
granola 125, 144
gravy (see sauce) 89
gremolata 195

H

heat transfer,
  conduction 179, 180
  convection 179, 180
  radiation 179, 180
homogeneity 37
humidity 178, 180, 242
hydrocolloids 220

I

instant pot, method 182, 183, 189, 194

K

knives (see utensils) 5, 6

L

leavening, use of
  baking powder 19, 20
  baking soda 19
  beer 32
leftover turkey tettrazini 217
lemon,
  garlic vinaigrette 41
  lemon meringue pie 140
lettuce (see salads) 35, 50
loaf, loaves
  blueberry cream cheese 31
  cinnamon 30
  classic quick bread 29
  Irish soda bread 32
  zucchini 30
lobster (see seafood) 168, 169

M

macaroni, mac and cheese 87
Maillard reaction (see browning) 20, 127, 216, 242
maple syrup 40
mayonnaise, classic 47
measurement, portions 10
meat,
  composition 172
  proteins, extracted 173
  science 172
meringue,
  aquafaba 146
  egg whites 147
  pavlova 146
  pie, lemon 140

mise en place 17
muffins,
  apple cinnamon 232
  apricot poppyseed 232
  banana chocolate chip 232
  berry 232
  berry granola 232
  blueberry brown sugar 232
  carrot walnut 232
  chocolate chip 232
  coconut lime 232
  cranberry orange 232
  jam 232
  lemon, poppyseed 232
  oatmeal, soft 231
  peanut apricot 232
  pear pecan 232
muscle,
  fibres 173
  protein, myofibrillar 173, 242
  structure 174
mussels,
  classic 171
  margarita 171
  marinara 171
  pesto cream 171
mustard baked eggs 132

N

Newton's laws,
  thermodynamics 178, 202
nuts,
  how to toast 54
  roasting 54

O

oatmeal,
  basic breakfast 229
  health benefits 220
  oat milk 230
oats 144, 220, 222, 229, 230, 231, 234, 237, 238
onions,
  caramelized (confit d'oignon) 117
  french onion soup 70, 73
  green 49, 52, 59, 79, 94, 95
  red 51
oranges 220

P

parmesan 48, 73, 86, 87, 114, 131
Parmigiano Reggiano 53, 169
partridgeberry,
  and apple jam 228
pastry,
  choux 138
  cream 138
  pie crust 140
pastry cream 138, 139
pavlova 146
pectin,
  also, pulp and peel of citrus fruit 220
poaching liquid 156
pork,
  chops with brandy cream sauce 192
  classic pulled 181
  fat (scrunchions) 160
  osso bucco 194
  shanks, pork (veal) 194
  shoulder 182
post-mortem 151, 243
potato(es),
  boiled 102
  duchesse 107
  frites 105
  mini pommes anna 113
  pancakes 102
  salad, classic 60
  salad, warm German 61
  scalloped 111
  shepherd's pie 107
  types of 99
  vichyssoise (potato leek soup) 81
  wedges 105
pot roast (see boeuf bourguignon) 190
poultry, temperature 7
prime rib 186, 187
protein-water matrix 198
pulling meat (see pork, rabbit) 183, 188

R

rabbit stew en croute 188
red peppers, how to roast 80
relish, million dollar 163
rice, basmati 204
rigor mortis 151, 152, 173, 243
roux (see sauce base) 83

S

salad,
  bean 58
  Cobb with mustard vinaigrette 57
  composing a 50
  easy Caesar 53
  gourmet tangy bean 59
  orange and almond 52
  pear and roasted pecan 54
  potato 60
  roasted beet and goat cheese 55
  simple green 51
  warm German potato 61
salmon (see fish) 152, 155
sauce,
  base: making a roux 83
  béchamel 84
  butter chicken 204
  cheese (Mornay) 86
  curry 86
  demi-glace 93
  espagnole 92
  garlic lemon cream 158
  gravy 89
  hollandaise 133
  making a sauce 82
  mother (French cuisine) 82
  mushroom cream 86
  velouté 89
sauce, toffee 225
scale, how to use 10, 11
scallops 63, 90, 158
science terms
  amylase 230
  amylopectin 64, 99
  amylose 64, 99
  beta glucan 220
  cellular 97, 98
  cellulose 78, 241
  coalescing 45
  collagen, as connective tissue 64, 173, 174, 179
  collagen as gelatin, gelatinous 64
  diffusion 178, 241
  dispersed,
    dispersion 35, 37, 45
  emulsification 35, 37, 83, 125, 198
  emulsifiers, 37
  emulsions 35, 45
  energy, 177, 178
  equilibrium, 178
  fatty tissue as a connective matrix 175
  gel networks 124
  gluten 20
  glycogen 173, 242
  heat transfer, 178, 180
  homogeneity 37
  hydrocolloids 220
  Maillard reaction 20, 127, 216, 242
  meat,
    composition 172
    proteins, extracted 173
    science 172
  muscle,
    fibres 173
    protein, myofibrillar structure 173
  Newton's law of thermodynamics 178, 202
  protein-water matrix 198
  rigor mortis 151, 152, 173, 243
  starches as carbohydrate 64
  water-holding capabilities 83, 173, 198, 220, 229, 242
scones (see biscuits) 26
scrunchions (see fat, pork) 160
seafood, (also see fish)
  authentic shrimp creole 94
  baked stuffed squid 167
  bouillabaisse 75
  chowder 63, 90
  lobster linguine tutto mare 168
  lobster tail, butterfly preparation 169
  seared scallops with garlic lemon cream sauce 158
  skate, *Raga radiata* 63
seasoning, creole 94
shellfish, temperature 7
shepherd's pie 107
shrimp (see seafood) 94, 152, 153
six sigma 4
sizing,
  diced 100
  julienne 50, 51, 59
  wedges 100
soup,
  bouillabaisse 75
  broth 70
  classic chicken noodle 71
  classic potato and leek 81

   creamy vegetable 78
   French onion with cheddar and ale 73
   pumpkin ginger orange 79
   roasted red pepper 80
   vichyssoise 81
spinach,
   bacon breakfast frittata 131
   creamed 114
squid (see seafood) 167
starches,
   as carbohydrate 64
   corn 64
stew,
   beef and barley 185
   bouillabaisse 75
   rabbit en croute 188
sticky toffee pudding 224
stock,
   beef 69
   chicken 67
   fish 68
stuffing,
   onion pecan 202
   traditional bread 167
   traditional down-home stuffing 211
super-juicy burgers 200

T

tarragon 41
temperature conversion 13
temperatures,
   chilling, cooling 7
   cooking, 7
tenderization 179
tenting, with foil 216
thermometer 6
thickening 64
timers, using 14
tomato(es),
   cherry 51, 57, 115, 169
   confit 115
turkey,
   maple sage anytime 215
   preparation 215

U

utensils, kitchen tools
   knives 5
   peeler, carrot 8
   strainer 8
   tongs 8
   whisks 8

V

vegetable medley,
   broccoli, brussels sprouts, and cabbage veggie bake 121
   roasted mixed vegetables 120
vegetables, root 64, 243
velouté, (see sauces) 89
vinaigrette,
   classic 39
   honey Dijon 40
   maple balsamic 40
   mustard 57
   sweet Asian 42
   tarragon 41
vinaigrettes 37
vinegar,
   balsamic 40

W

water-holding capabilities 83, 173, 198, 220, 229, 242

Z

zucchini 30, 50, 120

ABOUT THE AUTHOR

COLLEEN HISCOCK, B.SC. FOOD SCIENTIST AND CHEF

Chef Colleen Hiscock says that most everyone in Canada has eaten her food. That's because, as a food scientist, Colleen has spent thirty years developing food products for brands such as Maple Leaf Foods, Subway, President's Choice, Ziggy's, Compliments, Kirkland Signature, Sysco, Lilydale, Janes and many more. She's worked for some of largest meat and poultry manufacturers in Canada. During this time, she held highly technical roles as well as leadership positions that combined her mastery of food processing with in-depth business know how.

Colleen holds a bachelor's degree in Biochemistry (Food Science) from Memorial University, a culinary certificate in haute cuisine from George Brown College in Toronto, a Six Sigma Black Belt from Maple Leaf Foods, and an Interprovincial Red Seal in the culinary field.

She continues to share her expertise in commercializing farm-to-table food product ideas for the agriculture and food industries, and finds opportunities to add value to residual waste materials and for organizations that want to streamline production, enhance quality, and reduce cost.

Colleen is also co-owner and the executive chef at Java Jack's Restaurant & Gallery, a local fixture in wild and rugged Rocky Harbour, Newfoundland since 2000. She and her husband, Leslie, own and operate Java Jack's Bed & Breakfast just up the road from the restaurant. In a typical season, the restaurant serves some 12,000+ guests and the four-room B&B hosts more than 500 guests in a season. The restaurant provides breakfast to the B&B – a decadent and unmatched breakfast that is known as the best in the area.

Java Jack's boasts a customized line of delicious gourmet preserves and "dump & go" bakery mixes that Colleen developed, lovely coffee gift sets, and handmade items that appeal to both adults and children. These items are sold in many gift retailers in Canada and online at www.javajacks.ca/shop.

Building on her passion for food, Colleen started an online "Supper Club," where she demonstrates through online "webisodes" how to make and serve gourmet meals. Viewers cook along with Chef Colleen as she shares her favourite recipes and talks about the science behind everything from preparing burgers and making hollandaise to a wide variety of entrees and desserts. Each meal that Colleen and her viewers prepare includes some sort of culinary twist for the home cooks. These webisodes are available to view on YouTube @ ChefColleenHiscock.

The original Supper Club concept has been expanded into an in-person Supper Club event for guests to sip wine, watch the sun set, ask Colleen questions about food and cooking, and then enjoy a gourmet meal prepared by Colleen. For bookings, check out the website www.supperclubnl.ca.

Like Magic! is Colleen's first cookbook. She was driven by a desire to share her experience and knowledge with others so they too can understand the science behind food and cooking and make great food again and again and again.

Colleen and her husband live in Rocky Harbour, NL in the heart of Gros Morne National Park. There, she spends free time hiking, biking, swimming in summer, and cross-country skiing in the winter.

ACKNOWLEDGEMENTS

Big thank yous to my brother, Matthew, and my sister-in-law, Kelly, for helping me home in and providing clarity for the overall goal of the book. I always appreciate your insightful input and editing. To my best friend, Kimberley Hogan, who is always ready to help with anything I might be up to and consistently goes the extra mile on idea generation. And she makes it fun. Kimberley pulled out her collection of cookbooks and showed me which photographs enticed her to try out the recipe. She listened to me go on and on and on, getting it done.

A huge shout-out to Sally LeDrew for standing in at the last minute and taking on all the photography. After our first icebreaker photoshoot on October 2020, I knew we were going to create an amazing cookbook when our first "collection of" page appeared as the first page of every chapter with the building blocks of salads. That photo became the template for the entire project. It was easy to collaborate with Sally on the food, the shot, and the dish, and in the meantime, I gained a new friend.

Thank you all of you who let me borrow your dishes, plates, glasses and pots as props (Caroline Bugden, Julia Halfyard) and for your help with preparing several dishes and ingredients for the photoshoot (Pauline Ellsworth, Mandy Payne, Judy Decker, Darlene Hancock, Corey Greene). A thank you bouquet to the Gros Morne Farm and Market (Laurie Haycock, Boyd Maynard) for providing much of the vegetables and garnishes straight from the garden. I know that you hand selected them especially for this book.

I borrowed some lovely serving dishes and a few family heirlooms from my late mother-in-law, Lucy Hillier, before her passing. I know she would have been pleased with how I have honoured her by proudly showcasing her wares along with one of her recipes. Cooking was our special connection.

Thank you, Mommy, for letting me borrow anything in the house for the photoshoot. You always have beautiful dinnerware and accessories.

Susan Anderson (More than Words) for editing and establishing the format of one chapter for all the others and being able to disseminate the science ideas here in this book into concise parcels for easier reading. Audrey Heutzenroeder for your candid input, the hours of proofreading, your editing and suggestions and marketing support. I can't wait to see what will happen when we get to market. Joann Goosney, for the inspiration for the chef wizard illustration. My daughter, Heather, for various literature articles and editing all the references. My niece, Jamie, for a summer's work of creating the "how to" videos.

ONE LAST WORD...

You can purchase Java Jack's products at many gift stores in Newfoundland and throughout the rest of Canada. These products are at local regional grocers and on the shelves at Java Jack's Restaurant & Gallery in season. As well, you can purchase them on line at www.javajacks.ca/shop or at www.chefcolleen.ca and have them shipped directly to you.